THE
....
COMPACT
....
GUIDE
....

T0186942

D N A

OTHER TITLES IN THE COMPACT GUIDE SERIES

Winston Churchill
The Cold War
The Elements
Queen Elizabeth II
The Napoleonic Wars

THIS IS AN ANDRE DEUTSCH BOOK

Text © Andre Deutsch 2017
Design © Andre Deutsch 2019

Originally published in 2017 under the title *How to Code a Human*.

This edition published in 2019 by Andre Deutsch
A division of the Carlton Publishing Group
20 Mortimer Street
London
W1T 3JW

Typeset by JCS Publishing Services Ltd

Printed in Italy
All rights reserved
A CIP catalogue for this book is available from the British Library

ISBN: 978-0-233-00592-8

THE
COMPACT
GUIDE

D N A

Kat Arney

ANDRE
DEUTSCH

CONTENTS

INTRODUCTION

From a single fertilized egg cell to a complex human being, the genetic information encoded in our DNA directs the processes of life.

It is obvious from looking at any family that people who are closely related tend to appear similar. None of us is a perfect clone of our mother or father, though, and there are even differences between identical twins. Scientists have puzzled over this for thousands of years, devising all kinds of explanations for how traits and characteristics are inherited from our parents. However, it is only in the past century that we have discovered how this information passes down the generations in the form of genes, written in our DNA. It has also become clear that the path from genes to person – or from genotype (genes responsible for a particular characteristic) to phenotype (the physical manifestation of inherited traits) – is not a straightforward one.

To elaborate further, an adult human body consists of millions upon millions of tiny cells – nearly 40 trillion according to some estimates – forming organs and tissues with distinct characteristics, shapes and functions. There are hundreds of different cell types, ranging from lung, liver and lymph nodes to bowel, bladder and brain. Yet they all come from one single cell, which is created when a sperm fertilizes an egg. This cell divides in half, creating two cells. These grow and divide again to make four cells that, in turn, divide again and again. Over time, cells in different parts of the growing embryo begin to specialize, eventually creating all the various types of cells needed to make a baby. However, this process does not stop there. As the human body develops further, cells continue to divide constantly, replacing damaged and worn-out cells and repairing wounds.

We now know that genes control all these processes and also respond to the environment around them – nature and nurture are therefore both very important. Scientists are only just beginning to find the answers to some really big questions, such as:

- How does a fertilized egg divide and specialize to make all the tissues of the body?
- Why does a liver cell stay in the liver and a brain cell remain in the brain?
- How do your genes make you the person you are?
- How are traits, characteristics and diseases inherited?

During the 1960s, scientists started to piece together the concept of the molecular gene, discovering that genes are stretches of DNA encoding containing specific instructions for cells to do something (usually to make a specific molecule, such as a protein). Around the same time, computers were becoming more sophisticated, so it was easy to make comparisons between the strings of chemical "letters" that make up DNA and the logical string of digits or commands within computer code. Following on from this was the idea that if we could just crack the code and read all our genes, then we would understand exactly how our cells and bodies work.

Over the past few decades, it has become clear that this view is far too simplistic. Instead of being like computer code, with tidy electric circuits, genes are more like recipes. They are living entities, full of constantly shifting molecules and with many options for flexibility, depending on the range of things a cell needs to manufacture. Somehow, in the midst of this biological hurly-burly, genes need to be switched on and off at the right time and in the right place to ensure that cells continue to function properly.

It is also important to note that there is no such thing as a gene "for" a trait, such as height or intelligence, or "for" a disease, like cancer. Genes are recipes for making molecules, and it is how all these molecules work together in our cells, body and brain – along with the environment around us – that make us who we are and determines our risk of all kinds of illnesses.

In this book we will mainly focus on examples of human genes and traits, but many of the fundamental principles we will encounter are relevant across a wide range of living organisms.

We will look closely at the human genome, discovering how genes work and the way in which the twisted double helix of DNA encodes the instructions for life. We will track the journey made by our genetic ancestors out of Africa and around the planet and see what traces of them remain in our genes. We will find out how our DNA makes us who we are, right from the very beginning when the egg and sperm meet, and how cells "remember" what they are supposed to be doing. We will uncover the genes that shape our bodies and build our brains, and discover what happens when things fail to work as they should. Finally, we will look at what the future might hold, for both our genes and the entire human species.

Kat Arney

CHAPTER ONE

MEET YOUR
GENOME

**The human genome is made up of
roughly 20,000 genes, encoded within
billions of "letters" of DNA. But what is
actually in there?**

L et us start with DNA itself. More formally known as deoxyribonucleic acid, DNA is a type of chemical known as a polymer and is made up of repeating sub-units called nucleotides. Each nucleotide consists of three components: a sugar (deoxyribose); a small group of atoms called a phosphate group; and one of four different chemicals known as bases. These are usually referred to by the initials "A", "C", "T" and "G", which stand for adenine, cytosine, thymine and guanine. Human beings have a staggering 2.2 metres (just over seven feet) of DNA in almost every body cell, broken up into 23 matching pairs of chromosomes (long strings of DNA), all packed into a structure smaller than the head of a pin.

You receive one of each chromosome pair from your mother and one from your father at the moment of fertilization, when the sperm meets the egg. Each egg and sperm cell contains three billion "letters" of DNA in its 23 chromosomes – known as a haploid human genome – which come together to create the 23 pairs of chromosomes, making a diploid (combined) genome of six billion letters. Twenty-two of these pairs are known as autosomes, while the 23rd pair is the sex chromosome – XX in females and XY in males. This complete set of DNA, which is virtually the same in each cell, is your genome – the DNA code that defines who you are.

Photo 51: Solving the Structure

During the early 1950s, many researchers raced to discover the structure of DNA. A decade earlier, the experiments of the Canadian molecular biologist Oswald Avery had shown that DNA was the molecule in our cells responsible for carrying genetic information, but nobody knew what it looked like or how it worked. To solve this puzzle, Maurice Wilkins and Rosalind Franklin at King's College London used a technique called X-ray crystallography. By shining X-ray beams through solid crystals of DNA, they were able to see the "shadows" of the atoms inside and work out important information about the kind of structure that might be formed by DNA.

This photograph – known as "Photo 51" – was taken by Raymond Gosling, who worked alongside Franklin as a PhD student. The regular pattern suggests that the structure is a regularly repeating double helix or twisted ladder. Wilkins showed the photograph to James Watson and Francis Crick at Cambridge University. They were building models out of wire and cardboard to try to visualize the three-dimensional structure of DNA. It proved to be a vital piece of information, enabling Watson and Crick to determine how the nucleotides of DNA fit together to make the iconic double helix. In 1962, Crick, Watson and Wilkins won the Nobel Prize in Medicine for their discovery, but Franklin had sadly died four years earlier – for many years her contribution was less well known.

During the 1950s, the US/UK molecular biologist team James Watson and Francis Crick – together with Rosalind Franklin, an English chemist, and physicist Maurice Wilkins – discovered that DNA usually forms a double helix, like a twisted ladder. The sides of the ladder are long chains of sugar and phosphate, while the rungs are the bases. These bases can only combine in certain ways, with "A" always pairing with "T", and "C" always matching up with "G". Just as the order of letters in a recipe book determines the ingredients a cook needs for a pie and how to bake it, the order of bases in DNA (known as the sequence) carries important information about the biological recipes our cells need in order to live, multiply, specialize and even die. These recipes are our genes.

WHAT IS A GENE?

There are a few different definitions of a gene. At a molecular level, it is a specific sequence of nucleotides that usually encodes the instructions that tell a cell to make a particular protein (*see* Chapter 3). For example, a gene called "MYH2" contains the instructions used by muscle cells to build long, wiggly myosin (motor) proteins that help to generate force and make the body move. Taking a broader perspective, a gene is a unit of inheritance – information passed from parent to child imparting a specific characteristic or trait. So we might talk about blue eyes, increased cancer risk or even a talent for music being "in the genes".

Each approach has its advantages and disadvantages. The purely chemical definition of a gene – as a string of biological letters – is useful when scientists are studying how the molecules that build cells work together to keep the body going, but this has its limitations. Not all genes make proteins, and there are scientific arguments surrounding which types of DNA sequences count as "proper genes". In addition, up to 80 per cent of the genetic variations that have been linked to traits or diseases are not found in these protein-coding portions at all, but in the non-coding parts of DNA.

On the other hand, simply referring to everything that appears to pass from one generation to the next as being "a gene" is too simplistic.

There are very few characteristics that can be pinned down to a single gene or region of DNA. For example, hundreds of regions of DNA have been linked to intelligence or cancer, yet there is not (and never will be) a single "gene for cleverness" or a "cancer gene". As we discover more about how genes and the non-coding DNA around them work together, it is perhaps more helpful to think in terms of networks of genes and control switches working together, switching on and off at the right time and in the right place to build human beings and keep them healthy.

READING THE HUMAN GENOME

Geneticists have been using various techniques to work out the order of genes in organisms, such as fruit flies, for over a century. Throughout the 1970s onwards, they made much progress in pinning down the location of important human genes within chromosomes. However, to understand properly how genes and genomes work, they needed to map the underlying sequence of DNA. In 1977, two groups of scientists – one led by molecular biologists Allan Maxam and Walter Gilbert in the USA and the other by biochemist Fred Sanger in the UK – developed different techniques for reading the order of "letters" (bases) in DNA, known as sequencing. Eventually, Sanger's more simple method prevailed and formed the basis of the quicker and more advanced DNA sequencing technologies subsequently introduced. In fact, this type of DNA-reading approach has only recently been replaced with so-called "next generation" sequencing, with new techniques being developed and refined all the time.

By the 1980s, researchers were pushing for funding to sequence the entire human genome, even though reading just a few hundred letters was a painstaking process. Work began in earnest during the early 1990s, under the banner of the Human Genome Project (HGP) and directed initially by James Watson. Key organizations involved in the project included the National Institute of Health (NIH) and Department of Energy in the USA, as well as the UK-based Medical Research Council and the Wellcome Trust. Although slow at first, progress fortunately speeded up thanks to improvements in technology, including the development of automated sequencing machines and computer programmes for analyzing the data.

Pressure on the project intensified in 1998 with the establishment of the private company Celera by former NIH biologist Craig Venter. He aimed to sequence DNA using a different technique to that of the publicly funded teams. This initiated a race between the public and commercial projects to see which would be first to complete the whole human genome. It also led to arguments about the accuracy of the two approaches. Some doubted that Venter's new "shotgun sequencing" method – which involved blasting DNA into small fragments, sequencing them all and trying to match them up afterwards – was as good as the HGP's more methodical approach of sequencing long stretches of DNA from one end to the other. There were also real concerns that if data ended up in corporate hands, open access to it and any benefits to the global scientific community could be lost.

Not All Cells

It is often said that every cell in the body has exactly the same set of DNA, but this is not strictly true. As they mature, all mammals' red blood cells lose their DNA, possibly to allow more of the oxygen-carrying molecule haemoglobin to be packed inside. Some immune cells chop and change certain parts of their DNA that encode proteins called antibodies, which recognize and attack foreign invaders such as bacteria and viruses. This creates diversity in our immune system, enabling us to recognize a wider range of threats.

Furthermore, the DNA in cells gradually changes over time as it gets damaged (mutated). Chemicals in the environment, such as those in tobacco smoke, the ultraviolet rays in sunlight, and even the oxygen used by our own cells to generate energy, can all generate mutations, as can the process of copying DNA every time a cell divides to make new cells. In addition, there is no such thing as the "perfect human genome" – every human being has millions of small or large DNA changes that make him or her unique. Even identical twins, created when one fertilized egg splits into two and sharing the same DNA, may each have a few unique genetic differences.

Eventually, the two teams settled their differences. In February 2001, they published their results together, releasing the first draft of around 90 per cent of the human genome. This huge achievement brought together major advances in sequencing technology, DNA analysis and gene mapping. For the first time ever, we could gaze on our human genes – the biological recipes that make us who we are – and also marvel at how few there actually are.

JUNK OR GENES?

A few years before the completion of the human genome, a couple of researchers involved in the project set up a sweepstake for people to bet on the final number of genes in the body. Given that humans appear to make more than a hundred thousand different proteins across all the body's cell types, some estimates reached around 150,000. Others were much lower, at fewer than 30,000 genes, while the average guess was roughly 60,000. In the end, everyone was wrong. Although the winning guess was actually 25,947, the final number of genes in the draft human genome turned out to be 24,847.

In fact, it now appears that we have even fewer. Most estimates are between 20,000 and 22,000 genes, while some are even as low as 19,000. It is difficult to be precise because different researchers have their own definitions of what constitutes a gene at the level of the DNA sequence. After all, genes are biological entities and they do not always fall neatly into human-designed categories. It often causes surprise that such complex entities as human beings have so few genes, since this puts us on a level with much smaller animals, such as fruit flies and tiny nematode worms. Actually, many organisms have more genes than we do, with plants leading the league table.

It was also interesting to discover how little of the human genome is taken up with these 20,000 or so genes. Researchers have found that less than 2 per cent of DNA is composed of actual protein-coding sequences, and the function and purpose of the remaining 98 per cent is hotly debated in the world of genetics. It is sometimes referred to as "junk DNA", but non-coding DNA is a more accurate name. Some parts are structurally important for forming the middles and ends of chromosomes, known as centromeres and telomeres. Other regions are known to be important for gene regulation – switching genes on or off at the right time and the right place.

Around 50 per cent of our genome consists of repeated DNA sequences that originated from virus-like elements called retrotransposons, also known as jumping genes. These randomly hopped into the genome of our ancestors around four million years ago and have copied themselves many, many times since then. They cannot be very harmful to our DNA, otherwise they would not have persisted and multiplied over such a long time. Damaging genetic changes tend to be lost over time through evolution, or at least reduce to relatively low levels within populations. So much (if not all) of this repetitive DNA is likely to be junk. There are also broken or "dead" genes, arising from functional genes that were duplicated in the genome and picked up a damaging change along the way.

In 2012, a large project called ENCODE (the Encyclopedia of DNA Elements) published a number of high-profile research papers. These suggested that up to 80 per cent of the non-coding genome could be functional, based on whether certain proteins are stuck to DNA or whether sequences "read" to produce RNA (*see* Chapter 3). Nonetheless, simply finding evidence that interactions occur does not prove that these non-coding regions actually do anything useful or important within cells. However, the truth could lie elsewhere entirely. Based on comparisons between similar stretches of DNA in species ranging from humans to horses, mice to bushbabies, scientists in Oxford have calculated that less than a tenth of the human genome has a proper function. There's a vast difference between 80 per cent and 10 per cent, so, for now, the argument continues to rage.

It can be hard to grasp the idea that much of the genome is useless junk, but this is because it can be more comfortable to imagine design and order – especially in something as magnificent as the human race. Far from being neat lines of text or perfect computer code assembled in a sensible and methodical manner, our genome evolved over millions of years. It only needed to be good enough to make an organism that could survive in its environment and pass on its genes to the next generation. Sometimes pieces of junk DNA do become useful – for example, by becoming control switches or other gene regulatory elements – but, generally, the junk just stays there because it does not appear to be doing any harm.

Furthermore, other species have much more efficient genomes than we do, and they still build a perfectly functional organism. The Japanese puffer fish *Takifugu rubripes*, eaten as a delicacy and best known for harbouring a lethal paralyzing toxin in its organs, has a genome that is nearly eight times smaller than that of humans. Yet it has more or less the same set of genes and builds the same range of organs and tissues as humans.

A GENE BY ANY OTHER NAME

As we have already seen, the human genome is composed of around six billion bases, organized into 23 pairs of chromosomes and containing around 20,000–22,000 genes. Each gene needs some kind of identifying tag, so geneticists can be clear about what they are working on and communicate their results to other scientists. Some genes already have names, ranging from the sane and sensible (for example, MYH2, which stands for Myosin Heavy Chain 2) to the wild and wacky "Lunatic Fringe" and "Sonic Hedgehog".

There is a long tradition among researchers, especially those working with fruit flies and other smaller animals, of creating descriptive or strange names for new discoveries. However, faulty genes that cause human diseases are a different matter and funny or inappropriate names can be awkward or even offensive. To avoid such problems, a standardized naming system for human genes is used all over the world, with each gene identified through a handful or letters and numbers. The convention is to refer to genes in italic capitals – so the Sonic Hedgehog gene would be written as "*SHH*" – while the protein encoded by the gene (also known as Sonic Hedgehog) would be written "SHH".

The job of deciding the names of human genes falls to the Human Genome Organisation's Gene Nomenclature Committee (HGNC), coordinated by genetics expert Elspeth Bruford. She explains, "The aim of human gene symbols is that they should be used in all contexts, not just in scientific publications, presentations or discussions, but also in the media and by doctors in discussions with patients."

The HGNC team takes into account whether a gene has already been discovered and named in another organism, and also irons out any confusion arising if separate groups of scientists discover the same gene and name it differently.

"Ideally, we would name a human gene based on a known function of the gene product," says Bruford. "So, for example, if the gene encodes a protein with a known function then we would try to name it on the basis of that protein. If there is not a known function then we would start looking at how related it is to other genes that are already known or maybe have a known function. We would also look at versions of the gene in other species and see what was known about their functions. If there is nothing like that, then we would look at the structure of the

Junk in the Trunk

The term "junk DNA" was first in use by biologists as far back as the 1960s, although it became popularized through the work of the Japanese-American geneticist Susumu Ohno. In 1972, he published a paper entitled "So much 'junk' DNA in our genome" before anyone knew how many genes were in the human genome. He explained that humans have nearly 700 times more DNA than the gut bacteria *E. coli*, which has just over 4,000 genes. If the number of genes directly scales up with the size of the genome, then humans should have around three million genes, which seemed like an overly large number even then. Ohno also noticed that salamanders and lungfish have genomes more than 30 times bigger than our own, so they should have over a hundred million genes. Again, this seemed too far-fetched. Instead, he proposed that unimportant sequences of DNA tend to accumulate in genomes over time, even though they may not be useful. As Ohno put it, "The triumphs as well as failures of nature's past experiments appear to be contained in our genome."

protein encoded by the gene and see if it contains any specific regions, called protein domains or motifs, on which we can base a name."

Throughout the human genome there are also many stretches of DNA that look as if they should be genes and encode proteins, but scientists have not yet found proof that they do. These are known as ORFs, short for "open reading frames". Once researchers discover the function of a molecule encoded by an ORF, it can be given a proper name. So, having met the human genome, the next chapter deals where it came from.

CHAPTER TWO

OUR GENETIC

JOURNEY

**The journey of the human genome
started many thousands of years ago in
Africa. However, our family tree is not as
straightforward as it seems.**

One of the best-known images of evolution is the "March of Progress", which was created by the artist Rudolph Zallinger for a 1965 book on early humans. Although the picture is iconic, it is not factually correct. Species evolve gradually over millions of years, creating evolutionary trees that are branched and tangled rather than straightforward linear steps, as shown in the picture. That is just as true for human history and the diversity of life on Earth today. There are many examples of closely related species belonging to the same larger group. For example, there are different species of elephants, penguins, cats and so on. Nevertheless, modern humans – the species *Homo sapiens*, meaning "wise man" – are the only type of humans on the planet. So how did we get here? What were our ancestors like? How have they influenced our genes today?

HUMANS - THE EARLY YEARS

Life on Earth started nearly four billion years ago with simple single-celled organisms. The first fossils of bones that resemble those of modern humans were found in east Africa and date back around 200,000 years. These people are the oldest known examples of anatomically modern humans from our own species, although they were not the only hominins. In reality, the genetic and evolutionary journey of human-like species on this planet is both long and complicated.

Based on fossils that have been found so far, scientists believe that there are several main species of human creatures, known collectively as *Homo*. These include *Homo habilis, Homo erectus, Homo heidelbergensis, Homo neanderthalensis* (Neanderthals) and ourselves – *Homo sapiens*. The recent discovery of fossilized bones in the South African Rising Star cave system suggests that another one, *Homo naledi*, should be added to the list, although this is rather controversial. There are also other *Homo* fossils – often a few bones or even just fragments – that do not fit neatly into these groups and they have their own unique physical or genetic characteristics. An example of this is a partial finger bone belonging to a group known as the "Denisovans". Other interesting discoveries include tiny "Hobbits" found on the island of Flores (*see* page 20). It is not easy to decide whether these fossils represent a separate species or how best to fit them into the human family tree, and anthropologists continue to argue about this.

The most recent research suggests that our most ancient hominid (*see* page 14) ancestors were apes living in Africa about 10 million years ago. Over the next seven million years some became more like humans, standing upright on two legs and eventually learning to use simple stone tools. These animals include groups such as the chimp-like *Ardipithecus* ("ground apes", living roughly 4–5 million years ago), *Australopithecus* ("southern apes", alive between 2–4.5 million years ago) and *Paranthropus* (also known as "robust australopithecines", living around 1–3 million years ago). Then around two and half million years ago, in the Pleistocene era, a new species started to emerge – *Homo habilis*, or "handy man". Although there is some debate about whether these creatures truly belong in the *Homo* group or are more like the *Australopithecus*, they used stone tools to butcher and skin animals, and this gave them an advantage over the other primates around at the time.

The next big group (that we are aware of) emerged around 2 million years ago. *Homo erectus*, or "upright man", evolved in Africa and co-existed with *Homo habilis* for a few hundred thousand years, before being the first species to make the move out of the continent to explore other lands. They were clearly in existence for a very long time – possibly as recently as around 140,000 years ago. Fossils from these ancient humans have been found widely scattered across Asia. They were taller than previous hominin species, and there is also evidence that they used fire for cooking. Because we have no DNA from these fossils – as they are simply too old – we cannot be certain whether *Homo erectus* is truly our direct ancestor.

Homo erectus was followed by *Homo heidelbergensis*. This group is named after the town of Heidelberg, in Germany, where the first specimen was found in 1907. Experts believe that some *heidelbergensis* beings remained in Africa where they began their evolutionary journey towards becoming modern humans. It seems that others migrated to Europe and evolved into Neanderthals, eventually dying out around 200,000 years ago. Because species were changing and moving at different rates through disparate parts of the world, it is difficult to come to any real conclusion as to what was actually occurring at the time. Still, *Homo heidelbergensis* is our best guess at the last common ancestor of modern humans and Neanderthals, although a recent analysis of ancient DNA suggests that this might not be the case.

Sima de los Huesos, known as the "pit of the bones" and situated deep within the Atapuerca Mountains of northern Spain, contains fossilized bones from at least 28 individual ancient hominins, dating back roughly 430,000 years. Although scientists initially argued over whether or not they were *Homo heidelbergensis*, careful DNA analysis actually identified the bones as early Neanderthal. The finding also helps to pinpoint the likely timing of the split between modern humans and Neanderthals as being somewhere between 550,000 and 765,000 years ago. That is too far back to make *Homo heidelbergensis* our common ancestor, so scientists are now looking for another species to fill the gap. Fossils found in Spain from *Homo antecessor*, another closely related species, suggest that this could be the missing link. However, the mystery remains as it is unclear whether this group is a link in the evolutionary history of humans or simply a side branch on the increasingly complicated family tree.

ONE GIANT LEAP FOR HUMANKIND

The Omo River flows through the east African country of Ethiopia, eventually emptying into Lake Turkana in northern Kenya. Today, it winds through national parks and wildlife reserves, but 200,000 years ago its banks were home to what could be the oldest physical match to modern humans. In 1967, the Kenyan paleoanthropologist Richard Leakey and his team discovered two skulls and other bones. The best techniques available to them at the time suggested that these were roughly 130,000 years old – then the oldest known modern human fossils. Forty years later, new dating techniques pushed the timing back to around 195,000 years. Although these skeletons share many physical features with contemporary people, they are unlikely to have behaved in a modern, human-like way. Solid evidence of culture and technology within primitive societies does not appear until many tens of thousands of years later.

Hominid or hominin?

Classifying early apes and humans into different groups and species can become confusing. Broadly, scientists use the word "hominid" to refer to all ape-like species, and "hominin" for human-type species that have arisen since the split between apes and humans. The word "human" is also used to refer to all varieties of *Homo*, while "modern human" means our own, *Homo sapiens*.

While Europe and Asia were colonized by Neanderthals and other early hominins, our *Homo sapiens* ancestors stayed in Africa for at least 100,000 years. They grew in number and continued to evolve and develop as a species. Some groups may have left Africa, but it is difficult to be sure because the fossil record and ancient DNA contains so little solid trace of them. We do know that they were on the move between 60,000 and 70,000 years ago, probably in response to changing climate conditions in the region. During the first waves of migration early people travelled up into the Middle East and then out into the rest of the world. This was by no means an easy journey. Natural disasters and hostile weather may have pushed the migrating population to the brink of extinction several times. Today, with a world population of around 7.5 billion, it is hard to imagine that our species may once only have numbered tens of thousands.

Whatever happened, we do know that over the next 30,000 years or so these *Homo sapien* immigrants became the dominant species on the planet, while Neanderthals headed towards extinction. Nevertheless, the two groups continued to co-exist across Europe and Asia for thousands of years, and the latest research shows that there were many more similarities than was previously thought.

CLOSER THAN COUSINS?

Named after Germany's Neander Valley, where the first recognized fossils were discovered in 1856, Neanderthals lived between 30,000 and 300,000 years ago. Their position in the human family tree has been the subject of much scientific debate, but recent studies of ancient DNA from preserved bones (*see* page 19) has shed light on the relationship between Neanderthals and modern humans.

Neanderthals are commonly portrayed by the media and in museums as hairy, heavy-browed knuckle-draggers, but this is not entirely fair. We do not know for sure how their faces and bodies looked since soft tissues such as skin, hair and muscles are not preserved as fossils. Their skeletons and skulls show that they were generally shorter and stockier than modern humans, with large front teeth, prominent brow ridges and wide noses. Even though they had bigger brains on average than *Homo sapiens* – there is no strict relationship between brain size and intelligence – it is certainly possible that modern humans were smarter. However, Neanderthals were not complete savages and their culture was relatively sophisticated for the time. Archaeological digs have revealed that they wore jewellery and caught, butchered and cooked large animals for food.

In 2010, the Swedish geneticist Svante Pääbo and his team put together the first high-quality draft of the Neanderthal genome sequence. This was pulled together from samples taken from bones found in Croatia, Russia and Spain, as well as from the original Neanderthal in the Neander Valley. Their DNA shows that Neanderthals had brown eyes and paler skin than their African ancestors, but not as pale as today's Europeans, and some of them even had ginger hair. We also know they carried the same version of the gene "FOXP2" that we do, which is a gene associated with speech (*see* Chapter 12). Given that they also had a similar bone in the throat to modern humans, known as the hyoid bone, they could probably speak in some way.

Perhaps the most surprising result from this ancient DNA came when researchers lined up the Neanderthal genome alongside DNA from modern humans living in different parts of the world. They noticed that particular stretches of DNA in some groups of modern humans

Why Are There Still Monkeys?

Images like the "March of Progress" have promoted the idea that humans evolved from monkeys, and also raise the question: "If we came from monkeys, why are there still monkeys?" Based on the latest fossil evidence, scientists think that the split between early humans and apes started around 13 million years ago, but may have occurred over a further several million years. These early apes were certainly like monkeys, but definitely different from the chimps we see today. Just as the inhabitants, languages and borders of European countries did not stop shifting when the colonists went to America during the sixteenth century, we have diverged from our ape-like ancestors. In the same way, today's chimps, bonobos and gorillas have evolved from theirs.

exactly matched those of Neanderthals. Apart from certain tribes in Africa, it turns out that most people have at least some Neanderthal in them. On average, this works out at around 1–3 per cent of the human genome, although it appears that some people have more traces of these ancestral relatives than others. This must mean that early modern humans actually bred with Neanderthals during their period of co-existence. In the intervening millennia, most of their DNA has been lost from our genomes, but Neanderthal versions are still relatively common within areas of the human genome. Examples include genes involved in skin and hair colouring, the immune system, blood clotting, and the ability to break down fat and starch in the diet.

Many of these genetic variations would have been useful adaptations to life in Europe and Asia. Indeed, Neanderthals had already evolved to cope, so interbreeding was a quick way to acquire useful genes. However, thanks to today's environment and 21st-century lifestyles, some Neanderthal gene versions actually increase our risk of disease. One example of this concerns the Neanderthal version of a gene involved in blood clotting, which makes the blood stickier than the modern human version. This could have been useful if someone was wounded while out hunting, but today it increases the chance of developing a blood clot in an artery, causing a stroke.

Other Neanderthal DNA variations have been linked to a higher risk of depression, and new research suggests that they gave us a nasty version of the human papillomavirus, which causes cervical cancer. Intriguingly, two different genetic alterations that some people have inherited from our Neanderthal ancestors are associated with an increased chance of being addicted to smoking tobacco. Neanderthals did not smoke – tobacco was not even available that long ago – so, presumably, these genes served another, more useful, purpose at the time.

NOT JUST NEANDERTHALS

While we know that modern humans bred with Neanderthals and retained some of their genes, recent findings suggest that this also went the other way. DNA from the toe bone of a female Neanderthal from the Altai Mountains in Siberia reveals that she had sequences from *Homo sapiens* in her genome, which were put there some 50,000 years before she was born. This startling discovery suggests that modern humans and Neanderthals first met and mated tens of thousands of years before the major *Homo sapien* migration out of Africa. Perhaps this is the only remnant of an earlier brave wave of adventurers that died out, leaving their legacy in the DNA of the Neanderthals with whom they mated along the way. Genetic analysis also reveals that Neanderthals interbred with Denisovans – another extinct species of hominins living in the Mongolian Altai Mountains at the same time.

There are fragments in the Denisovan genome from earlier species, possibly the shadowy genetic remnants of *Homo erectus*. People from Fiji, Papua New Guinea and other nearby islands, as well as indigenous Australians, have around 4 per cent Denisovan DNA in their genomes. In addition, modern-day Tibetans have somehow managed to pick up a Denisovan version of a gene called "EPAS1", which helps them adapt to life at high altitude. More broadly, the indisputable fact that modern humans, Neanderthals and other species interbred – probably at more than one point in time – poses an important question: are they different species, or merely two "flavours" of the same one? A common view is that two groups of organisms are separate species if they cannot mate together to produce fertile offspring. However, the evidence that modern humans did breed successfully with Neanderthals is there to see in our genomes today.

Despite this, Neanderthals became extinct around 30,000 years ago. No one really knows why, although there are several theories. One concerns rapid climate change. At that time, the temperature could switch from freezing cold to pleasantly warm within a decade or so. Perhaps early *Homo sapiens* were better equipped to survive in these conditions than Neanderthals. Modern humans may have had better brains and certainly had smarter technology, including the ability to make clothes, tools and tents. They may even have been able to hunt with dogs, so, on all fronts, we could simply have been superior. It is also possible that we infected Neanderthals with diseases to which we ourselves had

Written in DNA

Although fossilized remains of our ancestors date back millions of years, it can be hard to determine how different species are related to each other just by comparing shapes and sizes. One way to uncover such relationships is by comparing DNA, in the same way that biologists work out evolutionary trees for species that are alive today. Unfortunately, the chemical structure of DNA breaks down over time, making it impossible to extract and read from very old fossils. Bacterial contamination within bones as they decompose is also a big problem, as is DNA that creeps in from modern-day humans and germs.

Probably the world's leading expert on the study of ancient DNA is Svante Pääbo, who now works at the Max Planck Institute for Evolutionary Anthropology in Leipzig, Germany. He and his team have developed techniques enabling them to analyze DNA from samples of fossilized bones and teeth dating back more than 400,000 years. The oldest modern human genome comes from a thigh bone found in Russia and is roughly 45,000 years old. Thanks to Pääbo's work, we now have genomes from early humans, as well as from Neanderthals and Denisovans. By comparing them with each other – and with people living around the world today – scientists can delve deeper into the complex relationships within human evolutionary history.

become resistant. Even though we bred with Neanderthals and swapped our genes with them, this was not on a grand scale. Studies of later Neanderthal genomes suggest that the populations involved were small and inbred, which would increase their susceptibility to infectious and genetic diseases.

As *Homo sapiens* is the only remaining hominin species, we cannot say for sure which of all the other *Homo* species are our direct relations, apart from Neanderthals and Denisovans. Interestingly, there seem to be traces of "ghostly" DNA in our genomes from extinct *Homo* species that are as yet unidentified. We may already have fossils from this group or this could be a totally new species altogether. Some theories also propose that modern humans evolved in several places in the world, although there is much more scientific evidence supporting the more popular "out of Africa" idea.

What becomes clear is that our history does not form a simple progression. There are overlaps and interbreeding between species, as well as extinctions and migrations all over the world. It is also important to remember that this all takes time – a *really* long time. Early humans did not wake up one morning and realize that they were a new species, or suddenly move thousands of miles away. Evolutionary changes and

Meet the Little People

The discovery of new human fossils in the Liang Bua cave on the remote Indonesian island of Flores captured the public imagination in 2004. The bones belonged to a peculiarly short adult female who was just a metre (slightly over three feet) tall. When the bones of further individuals were found a few years later, it became clear that her short stature was by no means unusual. So, although the species is officially known as *Homo floriensis*, they quickly became known by the nickname "Hobbits".

Some researchers think that *Homo floriensis* is a population of relatively modern humans with diseases or genetic traits that have made them smaller. Scientists have been unable to extract DNA from the Flores fossils but studies of their bones suggest that they are not directly related to us. It is possible that they evolved from an earlier wave of hominins – probably either *Homo habilis* or *Homo erectus* – that migrated from Africa to Asia hundreds of thousands of years ago and became small as a result of being trapped on their island. This phenomenon is known as "island dwarfism". Wherever *Homo floriensis* came from, they are no longer living on Flores, having died out around 50,000 years ago. Interestingly, this date corresponds with the probable arrival of *Homo sapiens* in the area, although researchers are still seeking proof that modern humans actually caused their extinction.

The discovery of the Flores people shows that there may be more extinct human species as yet undiscovered, although it is difficult to predict when and where these might appear. Fossils are usually only found by chance, although satellite mapping technology is helping to identify possible locations, while the activities of the world's mining and construction industries are also likely to turn up new fossils for scientists to study.

population movements take tens or hundreds of thousands of years, and there are no sharp dividing lines between species along the way. Dates, details and relationships will change as scientists gather additional data and find more fossils. All over the world they are piecing together the parts of this complex puzzle, and the picture changes as new evidence comes to light.

CHAPTER THREE

HOW DO GENES WORK?

Genes are the recipes in our cells, so we should get baking.

As any keen baker knows, the key to success lies in having an excellent recipe and following it correctly. Rather than making cakes or pies, our cells use genetic recipes, encoded within DNA, to cook up proteins. You might think of protein as the pink meat of a steak or the gloopy white of an egg, but proteins come in many guises and are the fundamental molecules that make up the body and keep it functioning properly. Thanks to many decades of research, scientists now know a great deal about how genes are read (a process called transcription) and how cells use these genetic instructions to build proteins (translation). However, before looking at that we should step back and find out what a gene actually looks like down at DNA level.

To start with, it should be noted that the words "read" and "reading" are mentioned in two different contexts throughout this book. In one sense, this means reading the order of letters (bases) in DNA – also known as DNA sequencing. The other refers to the molecular machinery inside a cell reading a gene when it is switched on, which is known as transcription. The word "letters" is also used to describe A, C, T and G – the four chemical bases that make up DNA (or A, C, G and U in the related molecule RNA).

ANATOMY OF A GENE

If you open a cookbook, it is usually fairly obvious where each recipe begins and ends. For example, it might start with the word "Ingredients", and finish with the instruction "Leave to cool and serve". By looking down the list of ingredients and reading the method, you can probably guess what you are cooking. If you make pastry and fill it with apples, spices and sugar, you will end up with a delicious apple pie. Rolling out sweet dough full of chocolate chips will result in cookies. In the same way, many human genes have a characteristic structure, so scientists can identify them and see what kind of protein they make – a task known as genome annotation. This is not always easy, but scientists have developed computer programmes designed to spot possible genes within the billions of base pairs that comprise our genome.

At the beginning of every gene is a stretch of DNA called a promoter. This acts as a kind of landing base for the molecular machinery responsible for reading genes and is usually around 250 DNA "letters" before the start of the gene itself. It can be difficult to spot a gene's promoter, as there are several different kinds and they do not all look the same. Generally, promoters consist of strings of short sequences that attract molecules, known as transcription factors, which help to recruit the gene-reading machine. There may also be other sequences even further away, called enhancers, which ensure that a gene is switched on in the right place and at the right time.

When identifying a generic recipe for a protein, scientists will look for long stretches of coding sequences. Every protein in the body's cells consists of one or more long strings of chemical building blocks called amino acids. The particular order of DNA bases – A, C, T and G – form a three-letter code, where each group of three stands for a particular amino acid or a signal to stop making the protein. Long, unbroken sets of triplets coding for amino acids, without any stop signals, are a useful sign that a particular stretch of DNA encodes a gene. There are also short stretches before and after the protein-coding recipe, known as untranslated regions, or UTRs. However, it is important to remember that not all genes look exactly like this. Some have unusual structures, while others encode messages that are not translated into proteins (non-coding RNAs). Some genes are very big, comprising many sub-sections known as exons (*see* page 27). There are also very small genes, referred to as small open reading frames or smORFs.

Cracking the Code

US scientist Marshall Nirenberg was one of the key people responsible for cracking the genetic code and discovering how certain combinations of DNA and RNA letters encode specific amino acids. Nirenberg and his postdoctoral researcher, German biochemist Heinrich Matthaei, created a kind of "cell in a test tube" that contained everything that was needed to translate RNA into protein. When they added RNA that was simply made of a long string of U letters, the system produced chains of the amino acid phenylalanine. The pair had worked out the first "word" in the genetic recipes of life, although they did not know how many letters were needed to specify each amino acid. Further research by Nirenberg and others revealed that the code is made of triplet combinations of letters and they eventually cracked the code for all 20 amino acids.

There is something else, too. DNA comprises two strands, each making up one side of the ladder. Just like stacking LEGO® bricks together, the sugars, bases and phosphate groups only fit together in one direction, meaning that each strand of the double helix is directional. Rather than talking about left and right, biologists refer to DNA as running 5' to 3' (five prime to three prime). The two strands line up in opposite directions in what is referred to as an anti-parallel arrangement. Furthermore, each strand is only read in one direction. This means that a gene can be encoded on either strand of DNA in different places in the genome. It is even possible for two different genes to be encoded in the same piece of DNA, but to be read in different directions from opposite strands.

READING THE RECIPE

The next step is to see how the instructions in genes are used to make proteins. It is helpful to imagine that the DNA in each cell is like a reference library full of recipe books. Each cell only has one set of DNA so it needs to be kept protected and safe inside a structure in the middle of the cell, called the nucleus. Just as books are not taken out of a reference library, DNA cannot venture outside the nucleus. You cannot cook a cake in the library and cells do not build proteins inside the nucleus. In the same way that you might photocopy the recipe from a book in the library and take it home to bake a cake in your kitchen, the cell makes a copy of the information in a gene and shuttles it out of the nucleus into the main body of the cell, the cytoplasm, where it can be used to make proteins. This process is known as transcription.

The molecular "photocopier" within cells is an enzyme called RNA polymerase and this comprises at least 12 proteins all working together. Many other proteins also need to be involved to ensure that the process works smoothly and that the right genes are turned on at the right time. To start transcription, RNA polymerase has to recognize and land on the promoter of a gene. It does this with the help of proteins called transcription factors, which attach to certain DNA sequences in promoters and enhancers near the start of a gene. Together, they create an attractive "landing site" for RNA polymerase, so it can start transcribing (more on this in Chapter 8).

Matters become further complicated as the DNA helix is untwisted and the two strands are pulled apart, breaking the weak bonds between each base and its pair. One side is the template – the actual genetic recipe – while the other is ignored. RNA polymerase works its way along the gene in a 3' to 5' direction, piecing together a molecule called messenger RNA (mRNA) that pairs up with the template. RNA (ribonucleic acid) is very similar to a single-stranded version of DNA and is made of a sugar called ribose (rather than the deoxyribose within DNA) as well as phosphate groups and bases.

Wobble, Wobble

There are 64 different three-letter combinations of A, C, G and U in RNA. Given that three of them are used as stop signals, cells should, in theory, need 61 different transfer RNAs (tRNAs) to cover all the different codons for 20 amino acids. In fact, there are only 45 or so in most organisms. Yet they can still recognize all the possible triplets in RNA thanks to something known as the "wobble hypothesis", first proposed by Francis Crick. He realized that the third letter in the anti-codon of a tRNA does not have to match up exactly to the one in the messenger RNA. The letter U or C in this position can be matched with a G in the tRNA, and an A or G in third place can match with a U. Some tRNAs also contain unusual letters, including one called inosine (I), which can match up with A, U or C in the third position in an RNA codon.

Importantly, RNA polymerase assembles letters that *pair* with the DNA template, rather than matching it exactly. For example, where RNA polymerase finds the letter G in the DNA, it will place a letter C in the corresponding RNA transcript, and vice versa. Where it sees the letter T, it will add an A. But rather than putting the letter T into RNA when it encounters an A in the template, the polymerase uses a very similar but slightly different chemical called uracil, known as U. So RNA is made up of the letters A, C, G and U, rather than the A, C, G and T of DNA.

Once RNA polymerase has reached the end of the gene it stops making RNA and falls off the DNA. It can then either return and re-read the gene again, making more RNA, or find another gene to work on. The RNA strand is released into the nucleus, but there are other steps to complete before it can be used to make a protein.

CUT AND PASTE, CAP AND TAIL

While food recipes are straightforward, unbroken instructions, recipes in our genes are more complicated. They consist of pieces of sensible, protein-coding information, known as exons, interspersed with introns, which are stretches of non-coding DNA. Imagine a recipe in which every

More Splicing, More Problems

Alternative splicing of messenger RNA is a powerful way to generate recipes for lots of different proteins from a limited number of genes. However, it can be risky, too. If there are any changes in the DNA or subsequent RNA sequence at the sites where splicing should take place, then inappropriate exons may be accidentally removed or included, or nonsense-filled introns left in. Some studies have suggested that anywhere between 15 and 60 per cent of DNA mutations that cause disease occur in splicing sites rather than protein-coding regions of genes, leading to the production of faulty or broken proteins.

Scientists have now found a clever way to manipulate splicing to treat the devastating genetic disease Duchenne muscular dystrophy (DMD) with a technique called exon skipping. People with this muscle-wasting illness have a fault in a gene called dystrophin, which is important for making muscles. Some versions of the faulty gene make RNA with a premature stop signal, which encodes a short version of dystrophin that cannot work properly. Adding specific, small fragments of DNA causes the defective exon to be spliced out, along with the unwanted stop signal. This creates a much longer protein, which can partly compensate for the missing dystrophin. The treatment, an injection known as Exondys 51 (or eteplirsen), has shown mixed results in clinical trials and is very expensive, but still offers hope to patients and their families.

few lines were broken up with nonsensical text. You would need to cut out all the nonsense before being able to read it properly, and something similar happens in cells. When RNA polymerase reads a gene it copies out everything, including all the introns and exons. These unwanted introns need to be removed from the resulting RNA and the remaining exons glued back together again.

This process is called splicing, and is carried out by a complex group of proteins known as the spliceosome. They either get to work alongside RNA polymerase as the RNA is being transcribed, or as soon as it is finished. Certain proteins within the spliceosome recognize specific RNA sequences at the beginning and end of each intron, and bring the two ends together to form a loop. The rest of the proteins in the spliceosome carry out the cutting and pasting, sticking the two exons together and releasing the intron as a lasso-shaped piece of RNA.

The process is made more complicated by the fact that nearly all our genes have many exons, and they can be spliced together in different ways. This is called alternative splicing. For example, the RNA from a gene with six exons could include all six of them. Or different combinations of two or more could be joined together, depending on the message that needs to be made. By switching up the splicing patterns in this way, cells can make a range of different RNA messages, each of which might carry the recipe to make a slightly different protein.

Over a hundred thousand different types of proteins are made throughout the body, so alternative splicing is an important way of increasing the number of possible proteins made in our cells from just 20,000 or so genes.

Once the RNA transcript has been spliced, two other processes must occur before it is ready to pass out of the nucleus and be translated to make a protein. The first is the addition of a special chemical called 7-methylguanylate, which is added onto the front (5') end of the RNA. This enables the RNA to be recognized by "cargo carrying" molecules that take it out of the nucleus. It also ensures that the RNA is translated efficiently and protects it from being accidentally broken down in the cell. The other step is to add a tail to the other (3') end of the messenger RNA. This is a long string of the letter A, repeated around 250 times,

Protein or RNA?

Not all genes encode proteins. Some are read into RNA, but it is the RNA itself that is used by the cell. One of the best-known examples of a non-coding RNA is called XIST (*see* Chapter 14), which is made from the X chromosome and is involved in shutting down one of the two X chromosomes in females. Some genes encode the RNA components of ribosomes, the tRNAs that deliver amino acids and other important molecules. There are even three different RNA polymerases for transcribing the varying types of RNA – RNA polymerase I, II and III. RNA polymerase II transcribes the messenger RNA that encodes all protein-coding genes, and certain non-coding RNAs, while RNA polymerases I and III transcribe the RNAs that make up ribosomes, as well as tRNAs and some other small RNAs.

known as a poly(A) tail. It is put on by a molecule called polyadenylate polymerase, and protects that end of the messenger RNA from being broken down, too. It also helps with protein manufacture, and is a signal to the translation machinery that it has reached the end of the message.

This, then, is the finished messenger RNA, or mRNA as it is known. It consists of a 5' cap, followed by an untranslated region, then the coding sequence made up of various exons glued together, another untranslated region and finally the poly(A) tail. It is now ready to leave the nucleus and head out into the cell's kitchen – the cytoplasm.

INTO THE KITCHEN

The process of converting the information in a strand of messenger RNA into a protein is known as translation. The cell's molecular "bakers" are blob-shaped clusters of proteins and RNA called ribosomes. Each ribosome is made up of two sub-units – one large and one small – that clamp together, with a string of mRNA feeding through between them. Millions of these molecules reside in every cell, and they are responsible for reading a strand of mRNA and assembling the right amino acid building blocks to make the appropriate protein.

As we have seen, the genetic code consists of groups of three letters. Each triplet is known as a codon, and each codon either corresponds to one of twenty amino acids or acts as a signal to tell the ribosome to stop making protein (see page 24). In order to assemble amino acids, the ribosome plucks amino acid-carrying molecules, called transfer RNA (tRNA), from the cytoplasm. Each type of tRNA is specific to one type of amino acid, and they all have a special loop structure called an anti-codon, which prominently displays three letters that pair up with the codon for that amino acid. For example, the anti-codon for the amino acid tryptophan is ACC, while the tRNAs that carry serine, another amino acid, could have anti-codons reading AGA, AGG, AGU or AGC.

Red Pen

Splicing, capping and tailing are not the only modifications that can be made to RNA. We now know that individual letters can be altered, and this phenomenon is known as RNA editing. Usually an A in messenger RNA is changed to I (inosine), and increasingly there are examples where this can alter important parts of the protein-coding sequence. More general editing also seems to occur across many human mRNAs. Evidence also shows that faulty editing is involved in motor neurone disease (also known as ALS or Lou Gehrig's disease) and other conditions. In addition, changes in the activity of ADAR, the enzyme responsible for editing, have been linked to cancer, viral infections, nerve disorders, autoimmune diseases and a number of other serious medical conditions.

The beginning of every mRNA contains an untranslated region, which does not code for a protein, so the ribosome's first task (after catching a piece of mRNA) is to know where the recipe actually starts. It scans along the RNA looking for the three letters AUG, which is the code for the amino acid methionine. Every protein in human cells starts with this building block, although the reason for this is not clear. The ribosome seizes the tRNA for methionine and holds on to it while it looks for the next triplet codon, pulling in that tRNA and amino acid too. These two building blocks are joined together, and the process carries on as the protein chain grows. If the ribosome comes across one of the three stop codons, it releases the protein and the RNA.

Once it has been read, the mRNA can be re-used by another ribosome to make the same protein, or it might be broken down. The freshly made protein will be folded up into the right shape, often with the help of special chaperone proteins, and may be modified in certain ways. Some proteins have chains of sugary molecules stuck onto them (glycosylation), while others may receive fatty molecules or small chemical "tags".

Many ribosomes float freely in the cytoplasm, churning out proteins that are used inside the cell. Others are embedded in a maze of membranes known as the endoplasmic reticulum, or ER, and make proteins that will be secreted outside the cell. The embedded ribosomes push their newly made proteins into the ER, where they are folded, modified, packaged into little blobs called vesicles and then sent on their way.

That covers the basics of what genes are and how they work. The next chapter will examine in more detail how DNA is copied inside cells and how harmful – or even useful – mistakes can creep in.

CHAPTER FOUR

UNDER

ATTACK !

**Human DNA is constantly being damaged,
with mistakes slipping in as cells divide.
Fortunately, the body employs plenty of
molecular repair experts.**

Every minute of every day the human body makes millions of new cells. This takes place, for example, in tissues that constantly renew themselves, such as the gut, blood and skin. New cells are created via division, as one cell splits in two. To achieve this, DNA and other important structures in the cell must all be copied then divided between the two "daughters". The process is by no means foolproof and errors can easily creep in. DNA is constantly under assault from the processes of life within the human body as well as external and environmental factors, such as chemicals and radiation. Spotting mistakes or damage and repairing these as quickly as possible is essential, otherwise errors in DNA can lead to such life-threatening diseases as cancer.

KEEP ON REPLICATING

Every time a human cell divides, each one of its 46 chromosomes needs to be duplicated, along with all the other material inside it. It may sound simple, but this process – called the cell cycle – must be carefully controlled so that each chromosome is completely and accurately copied once, and only once, before a cell splits.

The cell cycle is rather like a one-way loop round a racetrack comprising four sections, with checkpoints between each one. The first stretch is known as G1 or gap one. Here, a cell begins to copy all the important molecules and structures that need to be shared between the new cells, except for the chromosomes. Once this is done, the cell reaches the first checkpoint, and ensures that all is ready to proceed to the next stage.

The second stage of the process is S-phase (synthesis phase), when every part of the cell's DNA is copied – the task of a complex cluster of proteins called DNA polymerase. This peels apart the two strands of the DNA helix and moves along it, building a new matching strand for each side by pairing A with T and C with G. However, DNA polymerase only works in one direction along a strand of DNA – from 3' to 5' (see Chapter 3) – growing a new strand with the opposite arrangement. So only one strand of the DNA – the leading strand – is copied in a continuous, unbroken line running from 5' to 3'. The other – known as the lagging strand – has to loop around and be copied in short stretches, which are joined together by a special enzyme.

DNA replication begins at tens of thousands of specific places in the genome, known as replication origins. There are between 30,000 and 50,000 of these activated in each cell as it starts to copy its DNA, and different cell types will start copying DNA from particular replication origins at different times. It's not exactly clear why or how this happens, but it's probably related to the specific pattern of gene activity in that cell type, and how its DNA is organized.

At the end of S-phase, the cell reaches another checkpoint that ensures that every part of its DNA has been copied correctly. It then begins another gap phase, called G2. Here, the cell continues to grow rapidly, increasing in size and manufacturing many proteins to go into the new daughter cells. There is one more checkpoint to pass through, where the cell checks that there is no damage anywhere in the DNA. Collectively, G1, S-phase and G2 are known as interphase. The final step is mitosis. Up to this point, all the chromosomes have consisted of long strings of DNA packed into the nucleus like a tangled bowl of noodles. Now, each newly copied chromosome begins to twist up on itself, coiling around again and again to form a tightly packed sausage shape known as a chromatid. Importantly, the two copies of each chromosome stay in close contact, joined together in the middle via a structure called the centromere. Each pair is referred to as "sister chromatids", and together they form the classic X-shape that is often used to represent chromosomes.

Next, the sisters line up along the middle of the cell, attached through their centromeres to a complex arrangement of molecular "scaffolding" known as the spindle. By this point, the membrane around the nucleus has broken down, so the chromosomes are free to move around and organize themselves. The cell runs through a final set of checks to ensure that it has the correct number of chromosomes and that each chromatid is safely attached.

Doing the Twist

Chromosomes are made of long, twisty strands of DNA packed into a small space and can easily become tangled. This can cause major problems for DNA replication and cell division. Furthermore, unwinding DNA to be read or copied can lead to over-twisting further along the chromosome. To sort out the mess, cells use enzymes called topoisomerases that fall into two groups, known as type one and type two. Type one topoisomerases cut only one strand of the DNA helix and are used to relieve the stress on over-twisted DNA, while type two topoisomerases cut both strands to unravel more complicated knots. Some chemotherapy drugs for cancer work by blocking topoisomerases. Tumour cells end up with hopelessly tangled DNA and die because they cannot continue multiplying.

Finally, the spindle pulls on the chromatids, dragging one of each pair of sisters to opposite ends of the cell. The cell membrane pinches across the gap that is now left in the middle, sealing the separated sets of DNA into two new daughter cells. The chromosomes unwind themselves, the nucleus reforms and the cycle starts again with G1. The whole cycle takes roughly 24 hours for human cells, with G1 lasting around 11 hours, S-phase taking 8 hours, G2 being 4 hours and mitosis taking just a single hour to complete. However, cells do not necessarily multiply all the time. Those that do are more likely to be the specialized stem cells within most of the body's tissues. They continue dividing, producing either new stem cells or non-dividing cells. Those that have stopped going through the cell cycle enter a resting state known as quiescence, sometimes referred to as G0.

GENES AND GERM CELLS

Virtually all the cells in the human body are made through mitosis, with one key exception: eggs and sperm, technically known as gametes. These unusual cells are haploid, meaning they have half the amount of DNA compared to regular diploid body cells (also called somatic cells). Gametes have just one of each pair of chromosomes, so when egg and sperm come together during fertilization they make up the full set of 23 pairs. They are made from highly specialized germ cells, laid down very early on in development.

Gametes are formed through a modified version of the cell division process, known as meiosis. It begins just like mitosis, with all the DNA in the cell being copied and the chromosomes condensing down to form that characteristic X-shape made by two sister chromatids. Instead of all the copied chromosomes lining up in single file across the middle of the cell, each chromosome finds its pair and they line up together. So chromosome one pairs up with the other chromosome one, chromosome two with two and so on. When the cell divides, one of each pair goes into each new cell. Although each member of the pair originally came from either the mother or the father, the new cells receive a random assortment of maternal or paternal chromosomes. As each chromosome was previously duplicated, there is still too much DNA in each cell, and another round of cell division occurs, separating the sister chromatids. The final product is a set of four gametes, each of which have half the amount of DNA compared with a normal body cell.

The mixing and matching of chromosome pairs during meiosis is an important way of increasing genetic diversity within the next generation and creating new combinations of genes. There is another trick here, too. As the chromosome pairs line up together for the first stage of meiosis, they twist round each other. When they pull apart, tangled sections of DNA are broken and then glued back together again. Sometimes, a broken chromosome will be fixed using the same section of DNA from its matching pair. Swapping sequences like this, known as crossing over or recombination, creates entirely new and unique chromosomes that can be passed on. Crossing over usually occurs in relatively large "chunks", so genes close together on the chromosome are more likely to stay together through meiosis, and be inherited together, than sequences a long distance apart. This observation was used to make the first maps of chromosomes in fruit flies at the beginning of the twentieth century. It also explains why some versions of particular traits are always inherited together – a phenomenon known as linkage.

Crossing over can also cause problems. Female egg cells are made from germ cells very early on in development, while sperm are produced continually once a male hits puberty. In fact, the egg cell that you originally came from entered the first stage of meiosis when your own mother was still in her mother's womb. These cells lie dormant, with their chromosomes copied and paired, tangled and recombined. The trigger to go through the final division stages occurs when a woman ovulates, releasing an egg to be fertilized (or not). Being held for so long in this tangled state can make it difficult to separate pairs of chromosomes properly, especially the smaller ones. Sometimes an egg cell can end up with an extra copy of one chromosome. If this egg is fertilized, it will have three copies of that chromosome and that can lead to conditions such as Down syndrome.

The cause of Down syndrome is the presence of an extra copy of chromosome 21. In some cases, it has a relatively mild effect on the child,

Cycling Around

The different stages of the cell cycle are controlled by pairs of proteins known as cyclins and CDKs (cyclin dependent kinases). These were discovered by British scientists Paul Nurse and Tim Hunt, along with US researcher Leland Hartwell. All three shared the 2001 Nobel Prize in Physiology or Medicine for their findings. During his work with sea urchin embryos, Hunt noticed that a particular protein was created and destroyed in a regular, rhythmic pattern as the cells of the embryo began to divide. He called this new molecule cyclin. At the same time, Nurse and Hartwell were investigating strange-looking yeast cells that were having trouble dividing, because they were unusually small or unable to grow at all. These turned out to have faulty versions of their CDK genes. There are several types of cyclins and CDKs in human cells, and different combinations of a cyclin, plus a CDK, are responsible for moving a cell through the molecular checkpoints in the cell cycle and on to the next stage in the process. These molecules are extremely similar in all living cells with a nucleus (eukaryotes), from plants to pandas, making it a system that dates back to some of the earliest forms of life.

and some Down syndrome children grow up to be happy, high-functioning adults. Others can suffer severe disabilities and health problems, or the mother may actually miscarry. Nowadays, increasingly accurate tests are available during pregnancy for Down syndrome and other syndromes caused by the wrong number of chromosomes. People can now choose whether to continue the pregnancy – or not (see Chapter 5).

DANGER! DNA DAMAGE AHEAD

The human genome acts as the recipe collection for our cells, so it is important that mistakes (known as mutations) do not creep in. Although it might not matter if there are alterations in unimportant non-coding DNA, some changes can have a major impact on genetic instructions. These might be as simple as single letter typos. For example, changing a Z to an S in the word "liquidize" will not make that much difference to a recipe, as these are simply alternative spellings. However, altering a couple of letters in the word "tomatoes" results in "potatoes" – an entirely different ingredient that could have a profound effect on a recipe requiring tomatoes. Some mistakes are even more serious, with the equivalent of whole words or pages missing or jumbled up.

DNA mistakes cause problems for cells as they can no longer understand their recipes properly. In some cases, this might mean that a cell does not work as it should and stops making the right proteins. In more serious situations, changes to key genes controlling the cell cycle could make a cell start to multiply out of control, leading to cancer (see Chapter 16). Not all mistakes are negative, though. Some mutations can cause positive changes that benefit an individual, helping them adapt to their environment. If these are passed on down the generations they will shape and alter species over time. This is the basis of natural selection – the cornerstone of evolution.

Achieving Immortality

During the early 1960s, the American biologist Leonard Hayflick noticed that cells growing in the lab could only multiply a certain number of times – usually around 50 to 70 divisions – before they stopped dividing and eventually died. This so-called Hayflick limit occurs because the DNA and protein "caps" at the ends of each chromosome, known as telomeres, get shorter and shorter every time a cell divides. Like the plastic caps on the ends of shoelaces, telomeres prevent the ends of chromosomes from being nibbled away or accidentally joined together. Cells that need to grow and divide continually, such as stem cells, switch on an enzyme called telomerase, which can re-grow the telomeres to the correct length. Telomerase is switched off in regular cells, so when the telomeres get too short the cells stop growing and activate a kind of "suicide" process called apoptosis. This helps to protect us from cancer by preventing cells from multiplying when they should not. However, many cancer cells reactivate their telomerase and this enables them to continue growing out of control.

Mistakes can enter DNA in many different ways, and some problems begin with the fundamental business of life inside cells. Although DNA polymerase is very accurate when it copies DNA as cells divide, it is still not perfect. Occasionally, it puts in the wrong letters and can accidentally repeat or skip over sections of DNA. Another major source of mutation is damage caused by free radicals. These are highly reactive oxygen molecules generated as a by-product of the cell's energy-generating processes.

Additional sources of damage occur outside the body. Most people are familiar with the idea that smoking tobacco causes lung cancer. Chemicals produced by burning cigarettes enter the cells of the lungs and damage DNA. Something similar happens with air-polluting chemicals. Ultraviolet (UV) radiation from the sun causes two neighbouring DNA letters (two Ts or two Cs) to stick together. If not repaired correctly, this can lead to mutations and is a potent cause of skin cancer. X-rays are another source of DNA-damaging radiation, causing breaks in DNA strands, and many industrial or naturally occurring molecules can also cause mutations. Even types of food and drink, including red meat and alcohol, can harm DNA in the stomach, gut or food pipe and increase the risk of cancers developing there. Thanks to new DNA-sequencing techniques, scientists can now scan through genomes from tumours and healthy cells and see where DNA damage has occurred. They have discovered that certain sources of harm leave their own particular patterns of distinctive "scars" in the genome. This technology is still in its infancy, but in the future it could enable researchers to pinpoint the specific cause of someone's cancer.

Happily, some DNA damage is preventable. Giving up smoking, eating a healthy diet, enjoying the sun safely and reducing alcohol intake are all good ways to reduce exposure to potentially harmful agents. Unfortunately, some dangers are very difficult or simply impossible to avoid altogether. For example, even though free radicals are very destructive, we depend on oxygen to stay alive – human beings cannot prevent DNA damage by ceasing to breathe! On the positive side, cells are equipped with a handy repair team that is constantly on the lookout for problems.

THE GENE REPAIR TEAM

DNA damage alters the shape of the double helix, and this acts as a sign that something is wrong. Changes are detected by proteins that patrol the nucleus, checking that everything is in order. Some types of harm can also affect gene transcription or DNA replication, holding up RNA and DNA polymerase. This is another "red flag" to the cell's repair team, indicating that something is wrong. One of the key molecules responding to DNA damage is a protein called "p53", sometimes known as the "guardian of the genome". When damage is detected and depending on its severity, p53 directs the cell into one of three options: repair the damage if possible; stop multiplying and go into senescence (stop dividing); or die. In fact, the peeling skin of sunburn is the work of p53 triggering severely UV-damaged cells to die in order to prevent skin cancer.

There are several types of DNA damage, and these are each repaired by different "teams" of molecular repairers – clusters of proteins that work together to flag up and fix genetic errors. If both strands of the DNA helix are cut through this is known as a double-strand break, while a cut in just one strand is called a single-strand break. In this case, the remaining undamaged strand of DNA can be easily used as a guide to fill in the missing parts. Double-strand breaks are more difficult to fix, and are patched up in two different ways. The most accurate is homologous recombination, which relies on the fact that each chromosome has a matching pair. The repair machinery lines up the matching chromosome with the broken one and uses it as a template to repair the gap. If this is not possible, then a different set of repair proteins will just stick the two loose ends together via a procedure called non-homologous end-joining. That is fine as long as only one double-strand break is being fixed. If there are other broken bits of DNA in the nucleus there is a risk that the wrong pieces will be glued together, causing potentially disastrous genetic changes.

Some mutations can lead to mismatched pairs of DNA letters – for example, a C pairing up with an A rather than a G – which are also spotted and fixed. However, the repair molecules still have to work out which is meant to be the correct letter. In this case, replacing the A with a G would fix the mismatch, but so would swapping the C with a letter T. Only one of them will be correct. In some cases, it might not matter if a single letter is switched, but if it is in a crucial part of a gene or control region the consequences could be very serious.

Do Cell Phones Cause Cancer?

We know that some types of high-energy radiation – including X-rays and UV light from the sun – can damage DNA and cause cancer. Some people also worry that signals given off by cell phones (mobile phones) or masts might cause cancer too, especially brain tumours. Fortunately, the kinds of frequencies involved are too low in energy to damage DNA, and the exposure from masts and base stations is even smaller. Large studies involving many hundreds of thousands of people have also failed to show consistent, strong, reproducible links between cell-phone use and any type of cancer, including brain tumours. Their incidence has not significantly increased over the decades that these phones have been in widespread use. Nevertheless, it is difficult to rule out any connection completely and further large-scale research is currently underway. The European COSMOS trial, for example, is tracking 290,000 cell phone users for up to 30 years. As far as cell phones are concerned, it appears that the most serious health risk relates to accidents occurring because people are constantly being distracted by their phone.

DNA repair is absolutely fundamental for a healthy life. Without any repair systems at all our genes would be hopelessly damaged and scrambled within a very short time and we would be unlikely to survive long after birth, if we even made it that far. Problems with just one component of a repair team can lead to major health impacts. For example, faults in the genes encoding molecules involved in fixing single DNA letter mistakes can lead to accelerated ageing and very sensitive skin. People with a faulty version of p53 are at a much higher risk of developing several types of cancer – a condition known as Li–Fraumeni syndrome.

The "breast cancer" genes "BRCA1" and "BRCA2" are also involved in repairing DNA damage. Inheriting a broken version of either of these genes increases the risk of breast and ovarian cancer in women or breast and prostate cancer in men. In 2013, the actress Angelina Jolie made headlines by announcing that she had undergone a mastectomy (breast removal) to reduce her chances of developing breast cancer, having discovered that she carries a faulty version of BRCA1. More generally, everyone's risk of developing cancer increases with age. The body's repair mechanisms are imperfect, allowing mistakes that affect cells' ability to function properly or respond to environmental changes. This could be a key driver of ageing. If cells pick up mistakes in their DNA repair genes, they are likely to accumulate even more mutations, helping to fuel tumour growth.

Now we know how DNA is copied, damaged and repaired, and how variations and changes can creep in. The next chapter will look at how scientists study genes and what this can reveal about our health and the human body.

CHAPTER FIVE

W H O   D O   Y O U
T H I N K   Y O U   A R E ?

**From ancestry and Alzheimer's risk to
fingerprinting and forensics, how do
scientists uncover the information
hidden in our genes?**

Most people have seen TV crime programmes where a scientist holds up a striped X-ray film and points to a couple of lines that reveal the identity of the killer. They are also likely to have heard about paternity testing or the latest direct-to-consumer DNA tests that promise to reveal a person's hidden genetic ancestry, physical traits and risk of various conditions. Newspaper headlines frequently highlight the discovery of new genes for cancer, autism and more. Over the years, scientists have developed many techniques for studying the variations in genes, using them to identify individual people and the relationships between them, and drawing links between genes and appearance or health.

TAKING A FINGERPRINT

DNA fingerprinting was first developed during the 1980s by the British geneticist Sir Alec Jeffreys, at Leicester University. While searching for a means to track the inheritance of faulty, disease-causing genes through families, he accidentally hit on a vital method for telling the difference between people by looking at their DNA. To try to solve this problem, he and his team focused in on short, repeated stretches of DNA called variable number tandem repeats, or VNTRs. These form part of non-coding DNA, and there are nearly 30,000 of them scattered throughout the genome. As the name suggests, the length of each VNTR is highly variable, and every person will inherit a unique combination from their parents. Identical siblings, such as twins, will have the same pattern of VNTRs because they share the same genome. Closely related people have more similar patterns, while unrelated individuals are very unlikely to match.

Jeffreys and his team used molecular "scissors" called restriction enzymes, which cut specific sequences of DNA, to hack up a person's genome into short stretches. Because every person will have different length VNTRs, this means that all their fragments will be slightly different lengths too. The next step is to separate out all the individual fragments by size from shortest to longest (see box opposite) and then study the pattern, which looks a bit like a unique genetic barcode. Jeffreys quickly realized that he had hit on a brilliant idea. Simply comparing the patterns generated by two DNA samples, such as swabs from both a crime scene and a potential suspect, should reveal whether they are from the same person. It would also show whether two people are related, as there will be shared components in the pattern inherited across the generations in families, particularly between parents and their children.

Straightaway, this new technique – nicknamed genetic fingerprinting – generated a great deal of media interest. Interestingly, the first case to come to Jeffreys' laboratory did not involve either paternity testing or crime solving, but concerned a young boy who was due to be deported from the UK. By comparing the child's DNA to samples from the

Sorting by Size

To separate fragments of DNA according to their size, scientists use a method called gel electrophoresis. At the start, DNA mixtures are loaded into little pockets in a slab of gel that is placed in a tank and covered with a special liquid. The gel is made from a chemical called agarose, and its jelly-like texture allows fragments of DNA to travel through it. All it takes to get these moving is to pass an electric current through the tank. Because DNA is negatively charged, it will move towards the positive end of the tank. Smaller pieces of DNA can pass through the gel faster than larger ones and this sorts them out by size. Once this is complete, scientists reveal the pattern by staining it with dyes that stick to DNA. Alternatively, the DNA is transferred onto a thin membrane for further analysis, which might include a technique known as Southern blotting. Named after the UK molecular biologist Sir Edwin Southern, this method uses radioactively labelled probes to look for particular sequences of DNA.

woman who claimed to be his mother and her other three children, researchers proved beyond any doubt that he really was her son and that he should stay in the country. Thanks to this result, the laboratory was inundated with thousands of requests from desperate families in the same situation.

FROM LABORATORY TO COURTROOM

Over the past 30 years, DNA testing has become a widely accepted technique for determining guilt, innocence and family relationships. It is also used to identify victims of mass fatalities including natural disasters or terrorist attacks. The technology has moved on too, and scientists now use shorter stretches of DNA (known as short tandem repeats, or STRs) to generate a unique pattern. Studying shorter fragments means that fingerprinting can be done on older and degraded DNA samples that have already begun to break down. Jeffreys' original restriction enzyme method, which required relatively large amounts of DNA, has now been replaced by a technique called PCR. This makes many copies of the STRs from very small DNA samples, using just a few hair roots, skin cells or a tiny drop of blood to produce a viable result.

Glossy television shows and high-profile court cases have made people think that DNA fingerprinting has all the answers. Indeed, it can be very reliable if used correctly, but it is not perfect for several reasons. Because modern techniques are so sensitive, there is a risk of contamination if samples are not handled carefully. Errors and mix-ups can occur in testing, storage and analysis and sometimes it is difficult to interpret the results of a test conclusively. In addition, simply finding evidence of someone's DNA at a crime scene does not automatically prove guilt. We all shed DNA into the environment all the time, and there may be perfectly innocent explanations for the presence of someone's genes on the scene.

New Technique, Old Stories

Genetic fingerprinting has been used to find answers to historical cases such as paternity testing and proving the identities of long-dead people. The technique was used to identify the remains of the Russian royal family, killed in 1918 during the Russian civil war, as well as the body of infamous Nazi doctor Josef Mengele. Appropriately enough for a technique developed in Leicester, DNA fingerprinting was also used to prove that the skeleton found under a car park in the city was that of the fifteenth-century English king, Richard III, and that there might even have been a sex scandal somewhere in his family tree.

It is also virtually impossible to tell the difference between identical twins using regular fingerprinting techniques, although researchers are developing new methods that can distinguish them. Finally, there are serious ethical, privacy and security issues associated with gathering and storing genetic information about people on police or private databases. So, while there is no doubt that genetic fingerprinting has been a revolutionary technique, it is not a magic crime-solving silver bullet.

TRACING TREES

Another area where genetic technology has changed the world concerns our understanding of the links between traits, health conditions or diseases and modifications in specific genes. Since the 1800s, doctors and scientists have used family trees – known as pedigrees – to track how certain diseases pass down the generations. One of the most famous examples is the pedigree of Queen Victoria, who carried a genetic fault responsible for the serious blood-clotting disorder haemophilia on one of her X chromosomes (*see* Chapter 14). The other fully functional X chromosome found within Victoria herself balanced out this mistake, and within her female descendants who inherited that chromosome. However, in males, who have an X and Y chromosome rather than two Xs, there is nothing to compensate, so they will develop the disease. Eventually, haemophilia affected three generations of the British royal family.

As genetic knowledge and technology advanced over the course of the next century, scientists developed techniques for identifying regions of DNA – and eventually specific gene faults – that were responsible for several relatively rare inherited diseases. These range from cystic fibrosis and Huntington's disease to breast and bowel cancers that run strongly in families. Although this works in a limited number of monogenic (single gene) situations, the inheritance patterns of many common conditions and traits are less straightforward. This subject will be explored further in Chapter 6.

We know from studies that compare the inheritance of diseases and traits between identical twins (who share the same DNA and environment) and non-identical twins (who are as genetically alike as regular siblings and also share their environment) that many different factors have a significant genetic component. These include heart disease, cancer, diabetes, schizophrenia, autism, alcoholism, intelligence, height and weight. However, this does not definitively reveal what these genes actually are. It turns out that most conditions and characteristics are complex and polygenic, meaning that they are linked to many genes. Tracking these genes down and working out what they are doing has turned out to be a tricky task. Each trait is the result of proteins encoded by many versions of hundreds of genes, each of which only contributes a small amount to the final outcome. Unlike the straightforward pedigrees that can be drawn up for single-gene (Mendelian) diseases, people inherit a whole bunch of genetic variations that all add up to make them who they are.

IT'S A SNIP

Perhaps the biggest breakthrough in linking genetic variations to human traits and diseases occurred during the early years of the twenty-first century, with the development of genome-wide association studies (GWAS). These rely on technology to scan through the DNA of tens of thousands of people in search of single letter variations at hundreds of thousands of locations across the genome, scattered throughout genes and non-coding DNA. These variations are known as single nucleotide polymorphisms or SNPs (pronounced "snips"). So one person might have a letter A in a specific place, while someone else would have a G.

By comparing letters between thousands of people with a particular condition against those without it, scientists can narrow down certain SNPs associated with that disease. For most SNPs, there will be no association between the letter and having the condition, but some will show consistent links. For example, people with a letter T in a certain place might be much more likely to develop heart disease than people with a C in the same spot. By zooming in on the region of the genome where that SNP is located, scientists can search for genes or control switches that look like they might play a role in the disease and test them in the lab. An SNP for heart disease close to a gene involved in the growth or function of the heart is a good candidate for further study, while a gene that is only active in the skin is probably a red herring.

It is important to remember that SNPs could be in non-coding parts of the DNA as well as in genes themselves. Even within genes they could be in exons or introns. Furthermore, an SNP is not always a mistake responsible for causing a disease or trait. They are more like genetic markers that scientists use to pin down locations in the genome. Because of this, a single GWAS could reveal tens, or even hundreds, of SNPs that associate with a particular condition or characteristic. In fact, GWAS results are usually the source of newspaper headlines claiming that "Scientists have found 100 new genes for autism", or something similar. As discussed in Chapter 1, there is no such thing as a "gene for" – genes are the recipes used by cells to make proteins and RNA molecules.

Furthermore, the vast majority of these will have only a relatively small influence over the trait or condition with which they have been linked. For example, having a certain SNP might only increase someone's risk of bowel cancer by a few per cent and it is not guaranteed that they will develop the disease. However, if someone has many of the "bad" versions of SNPs, this can add up to a significant increase in risk. Nevertheless, this does not fully explain it. Twin studies have shown that genes contribute roughly half of the influence over a given trait in the population, on average, while the remainder is due to environmental factors. So, combining all the GWAS studies together for a particular characteristic or disease only explains a small fraction of the genetic component that should be there. So where is the rest of it?

Scientists have several theories about where to find this hidden "missing heritability", as it is known. One is to use even bigger groups of people, to try to capture less-common genetic variations. Another is to use a very narrowly defined description of what is being studied. A psychological disorder such as depression affects people in all sorts of ways, and there could be many genes at work. Restricting the study to a group of very similar people whose symptoms are all very similar increases the chances of finding stronger genetic associations.

Advances in technology mean that whole genome sequencing or exome sequencing (just reading the exons of genes and not the non-coding parts of the genome) is possible on a large scale. This will provide a lot more information about the genetic architecture of each individual and also better clues as to the links between genes and traits. Something else might be occurring, too – perhaps the inheritance of epigenetic marks (*see* Chapter 9) or molecules in eggs and sperm that could be passing on important information not encoded directly in the DNA. For now, this all remains a mystery.

GETTING PERSONAL

In recent years, there has been a great deal of interest in direct-to-consumer genetic tests offered by companies such as 23andMe, which is backed by the internet giant Google. Usually using the same kind of technology as GWAS, they look at whether someone is carrying SNPs that have been linked to a huge range of traits and diseases. For example, a 23andMe test can determine a person's risk of developing over 100 diseases, ranging from conditions such as cystic fibrosis, caused by a single inherited gene fault, to complex, multigenic disorders such as Alzheimer's or heart disease. It can also reveal if someone should like the taste of coriander (cilantro), what colour their eyes are likely to be and any potential sensitivity to the effects of several common pharmaceutical drugs.

These tests also claim to reveal details of genetic ancestry and the geographical origin of any ancestors. There are several aspects of this kind of analysis. For men, one test involves looking at the Y chromosome, which is passed down from father to son (see Chapter 14). There are limited types (haplogroups) of Y chromosome in the world, so two men who share the same Y haplogroup are more closely related than two that do not. However, this does not necessarily reveal anything else about their history or family relationship. The female version is mitochondrial DNA (see Chapter 14). This is the small amount of DNA found in the molecular "power factories" within the female egg cell and passed on from mothers to their female children. Again, there are a few broad groups of mitochondrial DNA, but this is not a specific test for ancestry.

More detailed ancestry tests analyze regions on other chromosomes, comparing them with information in databases of DNA variations from people all over the world. At the moment, this kind of test has limited value, being entirely dependent on the quality of the information contained in the databases. Furthermore, its reliability only stretches back a few generations. Because only half of each parent's DNA is passed on – and this is mixed up while eggs and sperm are made – you are bound to have many ancestors who have left no trace of their DNA in your

The Future of Fingerprinting

One day in the future, a genetic fingerprint might reveal much more than just a barcode. As more is discovered about how variations in genes affect the human body – and the cost of sequencing large stretches of the genome comes down – forensic scientists are now developing tools for making molecular photofits that suggest what a person might look like, based on their DNA.

The first and most obvious factor to look for is sex. Females will usually have two X sex chromosomes, while males have an X and a Y. However, someone's genetic sex does not always correlate to their outward appearance or gender (see Chapter 14). Next on the list are skin, hair and eye colour. Traits such as blue or brown eyes are relatively easy to work out, but green and other colours are still tricky. Hair colour can also be predicted, although that can be changed by a trip to the hairdresser. Scientists are starting to work out the relationship between certain gene variations and face shapes. Although it may not be specific enough to identify an individual – and face shape can change as people age or lose/gain weight – this approach could help the police focus on certain suspects rather than waste time with those who do not fit the profile at all.

Other traits, such as height and weight, are much more difficult to predict. Guessing age from DNA is another big problem, but scientists are developing new tests that can look at molecular "tags" that are put on certain genes as people age (see Chapter 9). Right now, this technology is still at an early stage, but it will become more accurate in time, as growing databases link genetic and physical characteristics.

genome today. So, although it can be interesting and fun to map your genetic ancestry, it may be more useful and informative to do more historical digging into your family tree instead. Adding to the confusion, humans have been migrating around the world for many thousands of years, matching and mating along the way. Nobody is genetically "pure", if you go back far enough.

Other, more specific, recreational genetic tests are currently on the market. One claims to design perfect skincare products based on genetic makeup, while another promises to predict a child's sporting ability from

Testing, Testing

Thanks to advances in genetic testing and reproductive technology,
families affected by serious or fatal conditions caused by a single faulty
gene can now choose to have preimplantation genetic diagnosis, or
PGD. First, in vitro fertilized (IVF) embryos are created in a laboratory,
using eggs and sperm from the mother and father, and allowed to grow
for several days into tiny balls of cells. A few cells are very carefully
taken from each embryo, then DNA is extracted and sequenced to
see which carries the faulty gene. Only a genetically normal embryo
is transplanted into the mother's womb and this should hopefully
grow into a healthy, bouncing baby. Another technique, called
preimplantation genetic screening (PGS), uses similar technology to
count the number of chromosomes in the cells of each embryo. This is
used to screen out embryos with the wrong number of chromosomes, a
condition that causes infertility and very serious health conditions.

It is also possible to detect some genetic conditions, such as Down
syndrome, later in pregnancy. Down syndrome is caused by the
existence of an extra copy of chromosome 21 (see Chapter 4). It is
usually diagnosed by testing the mother's blood for the presence of a
particular protein, and then confirmed by taking a sample of cells from
the amniotic fluid surrounding the growing fetus. However, this is
risky, potentially causing a miscarriage. Scientists have now developed
a test for DNA shed into the mother's blood from the placenta, which
has the same DNA as the fetus. Known as non-invasive prenatal testing
(NIPT), it can detect 98 per cent of cases of Down syndrome, as well as a
few other conditions caused by the presence of extra chromosomes.

their genes and suggest suitable activities for them. Although these are
based on scientific research, usually from SNPs identified in GWAS, they
are also of limited value. SNPs do not supply black-and-white yes/no
answers about complex traits, as there are so many genes involved and
the contribution of each one to the overall outcome is very small.

Professor Mike McNamee of Swansea University has looked into the
science and ethics behind these sports tests.

"These tests are looking for gene variations that predispose people
to endurance or sprint activities, for example. In the case of sprinting,

certain variations in a gene called 'ACTN3' are linked to sprinting speed. But there's a world-class Spanish long jumper who has a 'slower' version of the gene, so it's not an absolute requirement. And even if you do have that gene variation – and lots of us do – being a world-class sprinter requires many other things, like motivation, dedication and commitment, and there are no genetic tests for these traits," he says.

Genetic tests are likely to become more common over the coming years, especially for diagnosing, treating and monitoring diseases, such as cancer, in a more personalized and precise way (*see* Chapter 16). There are concerns about the privacy and security of large genetic databases that are now held by direct-to-consumer companies, as well as the potential risks of giving people potentially life-changing genetic information without proper advice and counselling. On the plus side, these products do get people excited and engaged with their genomes and offer an opportunity to share information with others, find new relatives and get involved in research. Nevertheless, there is still one major issue with any genetic test. The next chapter will look at the links between our genes and our physical makeup and health and show how these are not always straightforward.

CHAPTER SIX

PEOPLE ARE NOT PEAS

The basic rules of genetics were worked out over a century ago, thanks to a monk named Mendel and his pea plants. What do we know today about human traits and how they are inherited?

If you ever visit the city of Brno in the Czech Republic, be sure to visit St Thomas's Abbey and take a walk round the gardens. This is where the former abbot – a nineteenth-century monk named Gregor Mendel – worked out the fundamental principles of genetic inheritance. His work laid down the basic patterns that explain how characteristics and diseases are passed from one generation to the next, and underpinned many of the key genetic discoveries of the early twentieth century. Today, Mendel is recognized as the father of modern genetics, and the abbey where he lived and worked is now a museum in his honour. It has also become clear that life is far more messy and complicated than his neat flowerbeds. Although Mendel's laws are simple to understand, they do not actually work for most traits. Nonetheless, this is a good place to begin learning more about how genes and genetic variations interact with each other.

UNITS OF INHERITANCE

Thanks to distinctive inherited traits, such as flower and seed colour, Mendel was able to use pea plants to determine how different characteristics are passed on. He painstakingly transferred pollen between plants to control their reproduction, crossing purple-flowered and white-flowered plants together in specific ways and counting the number of offspring with each colour flower. To start with, he found some "true-bred" purple-flowered plants, which had only produced purple flowers for many generations. He then crossed these with some true-bred white-flowered plants and examined the flower colours of the resulting baby plants. Every single one of them produced purple flowers, with no white ones at all. Next, he bred these first-generation purple plants with each other. Intriguingly, while three-quarters of the resulting offspring had purple flowers, a quarter had pure-white blooms. After a lot of crossing and counting, Mendel concluded that the inheritance of flower colour and other traits, such as having green or yellow seeds, obeyed three basic rules.

First, he saw that each characteristic must be determined by two "units of inheritance", either one of which can be passed on unchanged to the offspring. He also spotted that the inheritance of one trait is separate from other traits – for example, inheriting red flowers does not always go hand in hand with green seeds – proving that they must be separate units. Finally, by measuring the proportion of offspring that inherit a particular trait, he realized that some versions of these units must be dominant, meaning that they are always expressed in the offspring, while others are recessive and only show up under certain circumstances. We now know that these units are genes, but nobody at the time understood that DNA was the genetic material in cells or indeed what genes were and how they worked.

LAYING DOWN THE LAWS

Mendel's laws neatly explain the genetics that lie beneath what he saw when he crossed his different-coloured pea plants together. Let us say that there are two versions (alleles) of the flower colour gene: P, which makes a purple pigment in the flower petals, and p, which has a change so that it does not make any functional pigment. Each pea plant has two copies of each gene – one from each parent – so pure-bred plants with purple flowers will have two P alleles (PP), and true-bred white flowers will have two ps (pp). If you cross a PP plant with a pp plant, every single one of the offspring will be Pp, because they inherit a P from one parent and a p from the other. In addition, because the P allele makes a purple pigment while the p version makes no pigment at all, the purple trait is dominant. Even just one P allele can make enough pigment to enable all these offspring to have purple flowers.

Matters become more interesting when these first-generation Pp plants are crossed together, as shown on page 74. Because each plant receives just one allele from each parent, there are four possible combinations. A quarter of the resulting offspring will be PP, half will be Pp or pP, and the remaining quarter will be pp. The PP plants will have purple flowers, as will the Pp and pP plants. However, a quarter will inherit both p alleles and have white petals. This produces the three-quarters to one-quarter (3:1) ratio that Mendel saw in his garden.

Mendel was keen to share his discovery with his scientific colleagues, so he wrote up his results in German and published them in a local research journal. Unlike today, when knowledge from libraries all over the world is available at the click of a mouse button, very few people heard about his remarkable findings and he died in 1884 without being recognized. Luckily, his work was rediscovered in the early twentieth century and became well known when the British biologist William Bateson translated it into English. Bateson also invented the word "genetics" to describe this exciting new area of science. The early geneticists took Mendel's ideas and developed them further, paving the way for decades of important discoveries about how genes are passed on. Eventually, this led to the discovery of the structure of DNA and the idea of the molecular gene – a stretch of DNA encoding the recipe for a protein.

On the other hand, some believed that Mendel's laws were there to be broken. Raphael Weldon, another British scientist, felt that Mendel's data looked almost too good to be true. He also examined the peas. Rather than seeing purely green or yellow peas, as Mendel claimed, Weldon saw a spectrum of colours ranging from dark green through greeny-yellow, yellowy-green and finally pale yellow. He concluded that the strict patterns of Mendelian inheritance do not often apply, and that most human characteristics are complex and variable, rather than controlled by the simple "units of inheritance" proposed by Mendel and his followers. Unfortunately, Weldon died in 1906 at the age of 46, just as Mendel's ideas were growing in popularity, thanks to Bateson and his colleagues. The more that is known about genes and how they work, the more scientists are realizing that Weldon was right.

MORE COMPLEXITY

Mendel was certainly correct in that we inherit two versions of each "unit" in the form of two copies of all of the 23 chromosomes – one from each parent – and these will usually have the same set of genes (with the exception of the X and Y sex chromosomes, *see* Chapter 14). However, this does not mean that the DNA sequence running the length of each pair is identical, only that the basic set of recipes is the same. There are likely to be plenty of differences in "spelling" within many genes, and even more in the non-coding DNA. There could also be even more significant changes, including missing or duplicated sections. In some cases, this might not make a difference to the protein that is made, while in others it could result in a small but subtle difference, or even a major change to its function. Furthermore, alterations in the non-coding DNA could affect how much of a protein is made, or exactly when and where the gene is switched on (*see* Chapter 8).

As we saw with Mendel's peas, each version of a gene is called an allele. You can think of these as being like "flavours", and some genes come in a very limited range. One example is a gene called "ABCC11", found on chromosome 16, which encodes a protein that transports molecules in and out of cells. There are two different alleles of the gene: one with a G in a particular position within the gene and another version with an A in the same place. This is a single nucleotide polymorphism or SNP (see Chapter 5). The G allele is dominant, and inheriting two G alleles or a G and an A version – one from each parent – means that you have wet, sticky earwax and smelly sweat. This version tends to be more common in people from Europe, Africa and South America. Receiving two A alleles is more common in Asia and results in dry, flaky earwax and less pronounced body odour. The inheritance pattern of the two alleles of the gene obeys Mendel's laws, and is the closest thing we have to a gene "for" earwax texture.

There are also a few hundred diseases caused by single gene faults and these do seem to follow Mendelian inheritance patterns, which is why they are known as Mendelian diseases. Examples include cystic fibrosis, caused by the inheritance of two faulty (recessive) copies of a gene called "CFTR". The healthy version of CFTR encodes a protein that forges a little channel on the surface of cells in organs such as the lungs, pancreas or skin, using this to shuttle salt in and out. If this channel is missing or fails to work properly, the imbalance in salt causes such symptoms as excessively salty sweat, sticky mucus in the lungs that interferes with breathing and a lack of digestive enzymes to break down food. Cystic fibrosis affects around one in every 2,500 babies, making it one of the more common Mendelian disorders. Other, very rare, inherited conditions also follow Mendel's rules, but this may be an oversimplification – some people can inherit two copies of a faulty gene yet be unaffected by the disease this should cause.

BITTERSWEET BIOLOGY

Scientists believed, mistakenly as it turns out, that several other traits fall into this neat Mendelian pattern. One is the ability to taste a chemical called phenylthiocarbamide (PTC). Some people think that PTC tastes extremely bitter, while others find it less so or cannot taste it at all. Researchers have identified several versions of a gene called "TAS2R38", which accounts for most of the variations involved in the ability to taste. There are two very common alleles, and the version responsible for PTC tasting is dominant over the one for non-tasting. However, with further alleles in existence, it is possible for two parents who cannot taste PTC to have a PTC-tasting child, and so this is not a simple Mendelian characteristic.

Examples of purely Mendelian traits, such as the "earwax gene", are very rare. Most genes have several different alleles, occurring at varying frequencies in people all over the world. Blood groups, which can be A, B, AB or O, are a good example of this type of trait. In this situation, both the A and B alleles are dominant, while O is recessive (see Chapter 13 for more about blood groups and "compatibility" genes). However, in many cases, it can be difficult to determine which alleles of a gene are dominant over others. It is rather like going into an ice-cream parlour and picking two scoops of any of the 20 flavours on offer. If you select two of chocolate, you will definitely have a chocolate ice cream, while the resulting flavour combinations from chocolate and banana, chocolate and strawberry or banana and strawberry will all taste different in their own way.

Even the inheritance of something that seems as simple as eye colour is quite complicated. This depends on the proportions of two different pigments – darker eumelanin and paler pheomelanin – in the cells making up the iris (the coloured part of the eye). Early studies of eye colour during the 1900s grouped everyone into three simple categories: blue/grey, green/hazel and brown. Scientists even drew up diagrams showing that blue and brown eyes fall into a Mendelian inheritance pattern, with brown being dominant over blue. They did not mention what happens with green-eyed people.

This idea is not really viable as eye colour does not follow straightforward inheritance patterns, and there are many hues and shades. Some people have coloured rings in their irises – mine are green with brown rings. Two blue-eyed parents can produce brown-eyed babies – something that should not happen if it is purely Mendelian. In fact, several genes contribute to eye colour, which come in a range of

In the Red

Although the idea that redheads have fiery tempers is not actually supported by science, we do know a fair amount about the genetic origins of flaming hair. Several genes are involved in colouration but red hair occurs because of one specific gene called "MC1R". This encodes a protein called the melanocortin 1 receptor, which is found in special pigment-producing cells called melanocytes. The colour of each person's skin and hair is mainly determined by the relative proportions of two pigments produced by their melanocytes – eumelanin (which is dark brown) and the paler pheomelanin.

People with brown or black hair and dark skin make a lot of eumelanin and tan easily. This acts as a form of natural sunscreen against damage caused by the sun's ultraviolet rays. Redhead variations in MC1R shift the production away from eumelanin and towards pheomelanin, resulting in red or blond hair. High levels of pheomelanin are also associated with having freckles and pale skin that burns easily in the sun. Unlike eumelanin, pheomelanin does not provide protection against UV rays, so redheaded people have a higher risk of sunburn and skin cancer. Intriguingly, analysis of ancient DNA from Neanderthal fossils (see Chapter 2) shows that our distant relatives also had a different variation in MC1R that would also have given them ginger hair. However, it remains unclear if their hair would have been the same shades of red that are found in humans today.

alleles. The most important gene is "OCA2", which makes a protein called P, found in the pigmented cells in the iris. Just one genetic variation seems to be responsible for blue eyes (see Chapter 8). Nonetheless, there are many ways of mixing and matching these eye-colour alleles to produce a wide variety of hues.

While it is still relatively easy to categorize eyes into broad colour groups, the same cannot be done with many other traits. Height, intelligence, weight and the risk of many diseases, such as diabetes, heart disease and cancer, all fall on a spectrum from high to low. These characteristics are the work of many genes interacting together, each of which could be one of many different alleles. Our unique characteristics come from thousands of genes all working together, creating all the proteins and RNAs that build the human body and keep it functioning. Variations in the coding parts of genes – as well as in the non-coding control switches – all have an influence on the final outcome, however small. Pinning down the interactions of all these components is a hugely complicated task and scientists are only just getting to grips with it.

ALL IN THE GENES?

Human beings are not simply a product of the genetic recipe book; the environment plays a major role too. Even identical twins, who share all their DNA, are not entirely identical. They have unique personalities and other traits, which result from non-genetic environmental effects. These cover many factors, from the conditions in the womb through to the food

Fitting into Your Genes

It is well known that weight is an inherited trait with around 70 per cent of the variation in weight between people being due to their genes. Environmental factors – including the availability of healthy food or an active lifestyle – also play an important role. Anyone will gain weight if they eat too much and do not burn off the energy, whatever their genes. Of course, there are some rare conditions where a single gene fault leads to someone being overweight, but for most people their weight is the result of the interaction between genes, environment and behaviour.

Several genes linked to obesity and weight have been identified through GWAS (see Chapter 5). Most of them encode proteins that are active in the brain, affecting the levels of molecules that drive hunger and control energy usage in the body. The strongest link is with a gene called "MC4R", which is closely related to the pigmentation gene MC1R. A recent small study showed that people with a faulty version of the gene prefer to eat fatty rather than sugary foods. This might have been useful in the distant past when food was scarce. Fat has more calories than sugar, so being driven to eat more would help someone to survive and thrive. Today, when most people have easy access to food and obesity is a major health issue, this is more of a problem than a benefit.

we eat, the air we breathe and everything we experience throughout an entire lifetime. Genes are highly responsive, and they work together with environmental factors to make us who we are and keep us healthy. For example, during a meal, genes that encode food-digesting enzymes need to be activated within the cells of the stomach so that your dinner can be broken down. If you get too hot, genes are activated to make proteins that protect the fragile structures in cells from becoming damaged.

Unpicking all the interactions between genes, control switches and the environment is very difficult. Thanks to advances in technology, scientists can now read the DNA from many thousands of people and match it up with patterns of gene activity, along with information about the body, its health and disease. As an example, the US government's Precision Medicine Initiative is recruiting a million volunteers for a study that collates extensive information about their genetics, lifestyle and health. In the future, this kind of research will help to explain how it all works together. Even so, every single person is a unique mix of nature (genes) and nurture (the environment), and we may never find out exactly what makes them into the distinctive individuals that they are.

The world has moved on a long way from Mendel and his ideas. Although some inherited conditions seem to obey his rules, it is not usually as simple as one gene for one trait or one broken gene leading to one disease. The next chapter looks at how researchers are sifting through whole genomes from many people and making some surprising discoveries.

G E N E T I C

S U P E R H E R O E S

**Although Superman and Wonder Woman
are fictional, real genetic superheroes walk
among us – and you might even be one.**

The link between genes and diseases is certainly complicated. For many common conditions, such as heart disease and diabetes, each person's risk is determined by a combination of many genetic variations working together with their environment and lifestyle. However, other illnesses do follow a simpler pattern of inheritance, with a single gene fault being passed down the generations. These are often called Mendelian diseases, as they generally follow Mendel's laws (*see* Chapter 6).

Some families affected by these conditions will decide to have genetic counselling and testing. Normally, a laboratory will look for DNA changes in the "usual suspects" – genes that are known to be linked to the disease – and compare the results between affected and unaffected family members. This does not always provide answers – the illness could be caused by a mistake in a gene or region of DNA that is not included in the test. Large-scale genetic studies are becoming more common now, thanks to a reduction in the price of DNA sequencing. Scientists are also gathering data from the genomes of thousands of healthy people in the general population, rather than just testing specific genes within families known to be affected by a particular disease. In doing so, they have discovered that all is not as straightforward as was once thought. It turns out that genetic superheroes are walking among us – and they may not even be aware of their hidden powers.

FINDING THE FIRST SUPERHERO

The first genetic superhero to be unmasked was Stephen Crohn. As a gay man living in New York and Los Angeles during the 1980s, he watched as many of his friends succumbed to a terrible disease that had started spreading through the community. This was eventually identified as AIDS (acquired immune deficiency syndrome), a progressive and eventually fatal illness caused by the human immunodeficiency virus or HIV. Despite being exposed to the virus, Crohn never caught it. Intrigued as to why he should be immune when so many others were not, he volunteered himself as a research subject. Scientists discovered that Crohn carried two copies of a rare version of a gene called "CCR5". The normal version of the gene encodes a protein that acts as a molecular door for the virus to enter immune cells, where it sets up an infection. To their surprise, the scientists found that Crohn's version of the gene effectively sealed this door shut, keeping the virus out.

Less than 1 per cent of people with European heritage have two copies of the resistant version of CCR5 and are largely immune to HIV infection (although this is not a complete guarantee of protection). A further 10 to 15 per cent have one copy. This does not make them immune to catching HIV, but it does reduce the chance of infection and slows down the progress of AIDS. Today, many life-saving drugs have changed HIV infection from a death sentence to a long-term condition with survival stretching into decades. One of these is maraviroc, also known by the brand names Selzentry or Celsentri. This drug, which sticks to the protein made by CCR5, preventing HIV from infecting immune cells, was developed as a direct result of finding Crohn's resistant gene.

Searching for Superheroes

Scientists at the Icahn School of Medicine at Mount Sinai are now going a step further in their search for genetic superheroes. They have set up the Resilience Project, an ambitious study aiming to recruit up to a million people and sift through their DNA to find people who carry "bad" gene variations, yet seem healthy. They are searching for people who have diseases in their family yet are unaffected themselves – good candidates for being superheroes. They are also interested in people affected by genetic conditions (who clearly are not resilient) or who are simply interested in getting involved and finding out more about their genome.

The hard work begins once researchers find people who have potentially harmful gene faults yet who appear to be fine. Scientists will have to work out why these lucky individuals have managed to avoid disease – whether that is because of their genes, lifestyle, environment or something else. Jason Bobe, who is leading the Resilience Project, describes this as searching for a "smoking airbag" – the opposite of a smoking gun. By finding out what is protecting these superheroes, Bobe and his team hope to find important new ways to prevent or treat many serious diseases in the future.

Stephen Crohn and others like him are genetic superheroes. Their CCR5 gene variation acts rather like a superpower, protecting them from a virus to which the vast majority of the population are vulnerable. Now, large-scale DNA sequencing is revealing that there are many more superheroes out there. Rather than being able to fight off the threat of viral infection, they are resistant to diseases caused by other faulty genes within their own cells.

IT'S A KNOCKOUT

In June 2014 a team of Dutch researchers, led by Professor Cisca Wijmenga at the University of Groningen, published some intriguing results. While reading the DNA of 250 families living in the Netherlands as part of a bigger project looking at the genetic makeup of people across the country, they discovered two people carrying two broken copies of a gene called "SERPINA1". Normally, this makes a protein that protects the delicate tubes and air sacs in the lungs. Without it – as is the case for people with two faulty versions of the gene – these intricate structures start to break down, eventually resulting in serious breathing problems by around 40 years of age. Despite this, both people taking part in Wijmenga's study were in their sixties with no obvious signs of lung disease – in other words, they were genetic superheroes. A staggering 177 people also appeared to be carrying genetic variations responsible for a bone condition called pseudoachondroplasia, yet they were in good health. Something elsewhere within their genome – or in their lifestyle or environment – was compensating for their faulty genes and keeping them well.

Two years later, a UK-based study of thousands of families – of mainly Pakistani origin living in London and Bradford – revealed a further 38 genetic superheroes, or "human knockouts", as the research team called them. These individuals were either carrying two faulty versions of genes linked to serious diseases or missing them altogether. Surprisingly, most of them were perfectly healthy. This population tends to favour marriage between blood relatives, so children are more likely to inherit two copies of a faulty gene. Although there is a higher rate of genetic diseases, this is not as great as might be predicted. As with the people in the Dutch study, other factors within their genes, environment or lifestyle are preventing them from developing a condition that they should otherwise have. The scientists behind this study, based at Queen Mary University in London, now aim to recruit a further 100,000 Pakistanis and Bangladeshis living in the city. The intention is to gain a better idea of what proportion of people in the population are genetic superheroes and to discover what is protecting them from disease.

An even bigger study from scientists at the Icahn School of Medicine at Mount Sinai, New York, scanned through databases containing genetic and health information from over half a million people and discovered 13 more superheroes. All had two copies of faulty genes that should have caused serious inherited conditions, including brain, bone, skin and auto-immune diseases. For example, based on their genetic makeup, three of the superheroes should have had cystic fibrosis – a classic Mendelian disease affecting the lungs and other organs. Yet they were apparently unaffected. This story made headlines around the world, but unlike Peter Parker, Diana Prince or Clark Kent – the alter-egos of Spiderman, Wonder Woman and Superman respectively – the true identities of these genetic superheroes will forever remain secret. This is because researchers did not have the right information or consent to track them down. That is a problem because it means that it is not possible to double-check the results of the study or determine whether any of these people have milder versions of these conditions – or even if they truly are superheroes.

SUPERHEROES TO THE RESCUE

Based on studies done so far, superheroes do exist all over the world. In fact, any one of us might have genetic superpowers – we may each carry up to 40 "bad" gene variations or faults, yet these are balanced out by the rest of our genetic makeup, environment or lifestyle. Interestingly, one country – Iceland – tends to produce many superheroes. Because its population is so small and it has been relatively isolated for a long time, the people of Iceland are more closely related to each other than in most other places. In fact, nearly 8 per cent of them carry two copies of genetic variations that should cause disease, but do not always exert their harmful effects. This phenomenon is not restricted to humans, either. Scientists have found "superdogs" that are protected against the canine version of the muscle-wasting disease Duchenne muscular dystrophy.

There is more to this work than just seeking out superheroes, however exciting that sounds. Just as Stephen Crohn's protective version of CCR5 led to the development of an important HIV drug, scientists hope that finding the genetic or environmental factors that protect these fortunate few will provide future treatments for those who are less lucky. It also indicates that we need to stop thinking about genetic diseases in such a black-and-white way. Even illnesses that were thought to be strictly Mendelian now seem to show a wide spectrum of symptoms in people who carry those gene variations, ranging from severely affected at one end through to apparently completely healthy at the other. This has important implications for genetic testing and counselling, as it shows that we cannot yet fully predict what someone will look like or how their health (known as their phenotype) will be, based on their genetic makeup (genotype).

Not All Bad

Scientists have had some impressive successes in their search for genetic superheroes. However, one big issue concerns the database of human mutations, which lists all the gene faults currently linked to disease. Cisca Wijmenga's study (*see* pages 66–67) showed that some versions that are meant to cause certain conditions are very common within the Dutch population. This does not seem right, as disease-causing variations should be rare. Instead, it is likely that there are mistakes in the database, so some of these supposed "bad" genes do not actually cause problems at all.

As another example, the majority of illnesses caused by mutated forms of proteins called prions, such as Creutzfeldt–Jakob disease (CJD), seem to be linked to inherited faults in the gene encoding the prion protein itself. When Professor Daniel MacArthur, a geneticist at MIT's Broad Institute, rifled through the DNA of more than half a million people – including 16,000 individuals with conditions caused by prions – he made a surprising discovery. Although some variations in the prion gene are definitely responsible for disease, others have been falsely accused and are not actually harmful. In addition, some "superhero" versions cause disease in some people but not in others.

FEELING REDUNDANT

There is another complication to take into account when determining how genetic faults or variations are linked to disease. We have already seen in Chapter 6 that many genes come in different "flavours" or alleles, and that there can be complex relationships between them in terms of dominance. It is also well known that humans inherit two copies of every gene – one from each parent, with the exception of the sex chromosomes (*see* Chapter 14). In many cases, if one allele is completely broken then the other one can make up for the loss, at least to some extent. If both copies are broken, then the expected outcome is some kind of illness or other condition. The existence of genetic superheroes shows that there must be other back-up systems at work to compensate for missing or faulty genes.

Since sequencing the human genome, scientists have discovered that many human genes are not unique. There may be two or more slightly different versions within the genome, each of which appear as two alleles, one on each member of a chromosome pair. These new versions arise because genes can get accidentally copied and repeated in the genome, usually because of mistakes in DNA copying or repair (*see* Chapter 4). Over

Superworms

By studying tiny nematode worms, scientists at the Centre for Genomic Regulation in Barcelona have discovered one possible explanation for genetic superheroes. These animals, known as *C. elegans*, can be bred to be genetically identical and they all share the same environment. So they should all respond in the same way to changes in their genes. Professor Ben Lehner and his team used genetic engineering technology to create a fault in one particular gene in their worms, which should have been fatal. However, only half of the worms died. When they broke two genes, 90 per cent of the worms died, yet 10 per cent still survived. Lehner thinks that random variations in the levels of transcription from other genes, or regions elsewhere in the genome, compensate for the missing genes in those superworms that survive, keeping them alive and wriggling.

time, these copies begin to evolve and become distinct from each other, often in quite subtle ways, and are referred to as being a gene family. In the human genome, for example, there are more than 30 genes for myosin, and these are active in muscle and other tissues. Some are used in particular ways in different parts of the body, but others are less specialized. So, if one member of a gene family is broken or faulty, then another one can come to the rescue. This biological back-up system is known as redundancy, and helps to explain why many of our 20,000 genes do not seem to be absolutely essential.

There are examples where scientists are trying to use redundancy as a way of treating disease. Humans and other mammals have a cluster of closely related genes called the beta-globin locus, which encode proteins that form part of haemoglobin – the iron-containing protein that shuttles oxygen around the body in our red blood cells. These genes all evolved from the duplication of one single gene millions of years ago in our evolutionary history and they are used at different points in life, from fetus to adult. People with the diseases beta-thalassaemia (or sickle cell anaemia) have two faulty copies of the adult version of beta-globin. They cannot make haemoglobin properly so they become ill. However, most people with a faulty adult beta-globin gene still have a perfectly normal version of beta-globin while in the womb – it has just been turned off as they develop and grow older. Researchers are now working on ways to switch this healthy fetal version of beta-globin back on, which could cure the illness.

Finally, it is important to remember that variations, changes and mutations to DNA do not necessarily have to be within genes themselves. They could be in the 98 per cent of the genome that does not code for proteins, affecting the control switches that turn our genes on and off at the right time and in the right place. The next chapter delves deeper into the genome to discover more about these switches and what happens when they go wrong.

CHAPTER EIGHT

TURN ME ON

**Genes need to be switched on at the right
time and in the right place to keep the
human body in optimum working order.**

The human body has around 20,000 protein-coding genes and these make up less than 2 per cent of the entire genome. Clearly, all these genes cannot all be switched on in every single cell, otherwise every cell would make an identical set of proteins and they would all be the same. Instead we have hundreds – if not thousands – of different types of cells, each doing their own particular job thanks to their own special set of proteins.

Some proteins – produced by so-called housekeeping genes – are needed within all of these different cell types, carrying out fundamental jobs such as making structures inside cells or generating energy. However, many genes are active just within specific groups of cells. They need to be carefully controlled (regulated) to ensure that they are switched on only as and when needed, and in the right combinations. The challenge associated with gene regulation goes right back to the very beginnings of life, mapping out the complex patterns of gene activity that direct the growth of a single fertilized egg cell into a baby. Diseases, such as cancer, can be caused by important genes being switched on or off at the wrong time, leaving cells to multiply out of control and grow into tumours.

Gene activity is controlled by "switches" within the non-coding DNA, which, as we have seen in Chapter 1, comprises the other 98 per cent of the human genome. Although humans may only have around 20,000 genes, we seem to have hundreds of thousands of switches, each of which acts on its own or in combination with others, to turn genes on at the right time in the right place. Intriguingly, around 80 per cent of the genetic changes (SNPs) that have been linked to diseases are found in this non-coding DNA, suggesting that alterations to the switches or other control elements are just as, or even more, important as changes within genes themselves. So, how do they work?

THROWING THE SWITCH

Usually known as regulatory elements or enhancers, control switches are short stretches of DNA composed of sequences of "letters" or bases. They act as docking sites for proteins called transcription factors, which attract RNA polymerase and the rest of the transcription machinery to a gene when it needs to be read (*see* Chapter 3). The physical shape of the DNA sequence itself determines the specific transcription factors that can bind there, and different factors have different affinities for particular sequences. For example, one transcription factor can only stick to the sequence "TACGTA", while another might be less picky, and be prepared to attach to a range of similar DNA sequences.

Enhancers typically consist of several transcription-factor docking sites strung together, one after the other. Transcription factors also like to bind to DNA co-operatively, meaning that if one has already settled down onto an enhancer, others are more likely to follow. Once sufficient transcription factors are stuck to an enhancer, this creates a kind of

Got Milk?

Around three-quarters of the world's adults – particularly in East Asia – are lactose intolerant and so are unable to drink milk or eat dairy products. For the remaining quarter, one change in a control switch is all it takes to make the difference. In most mammals – including those who are lactose intolerant – the gene encoding an enzyme called lactase, which breaks down sugar in milk, is switched off early in life. However, around 10,000 years ago, one of our ancestors in Europe was born with a change in a control switch near the lactase gene, meaning that this continues to be transcribed throughout adult life. In addition, other genetic changes within the non-coding DNA around the lactase gene in African and Middle Eastern populations occurred separately, but have the same effect in keeping the gene active. Wherever they arose, the gene changes associated with lactase persistence seem to have gone hand in hand with the spread of dairy farming in the area. Milk is a good source of protein and energy, so being able to farm cattle would have been a big advantage.

"landing platform" for RNA polymerase, directing it towards the start of the gene so it can commence transcription (*see* Chapter 3).

The specific arrangement of these sites, together with the combination of transcription factors in a particular cell type, determines whether a gene will be activated in any given cell. Although all cells have the same genes and switches – since they all have the same genome – different cell types will have distinct blends of transcription factors that activate only the genes that they need rather than those they do not. Because the DNA sequence is so important for creating transcription-factor binding sites, it is easy to see how any letter changes might affect how tightly the transcription factors can attach. This, in turn, will have an impact on how effectively they can activate transcription.

TRACKING BACKWARDS

If genes are activated in specific cell types by transcription-factor proteins sitting on their control switches, then a logical question would be: which particular switches on the genes make those transcription factors?

There are two ways in which this can occur. In the first instance, some genes need to be switched on in response to changes in the environment, inside or outside a cell. For example, genes encoding proteins that break down food and transport it around your body need to be activated when you begin eating. Molecules involved in DNA replication and cell division must be switched on when a cell receives signals telling it to split into two new cells. If you catch an infection, genes in the immune cells need to kick into action to make antibodies to fight it off. In these cases, cascades of chemical messages – known as signalling pathways – fire into the cell, activating dormant transcription factors by changing their shape or structure. These activated transcription factors can then go and seek out their enhancers, switching on the necessary genes to evoke the correct response.

Then there are genes that are always active in certain cell types. These include those encoding stretchy myosin proteins in muscles, oxygen-carrying haemoglobin in red blood cells, or detoxifying enzymes in the liver. In such cases, we need to go back and look at the history and origin of these cells. As a single fertilized egg develops into a fetus in the womb, the cells divide and specialize. Along the way, groups of cells turn on particular patterns of genes, pushing them towards a particular fate, such as becoming muscle, nerve, bone or guts. Eventually this leads to the development of all the organs and tissues that are needed by the body to function (see Chapter 11). Adult muscle cells make myosin because they contain muscle-specific transcription factors. This is because they came from earlier muscle precursor cells in the fetus that had switched on these genes. In turn, those fetal muscle precursor cells came from a less well-defined group of cells in the middle layer (mesoderm) of the embryo at an even earlier stage. Some of these mesoderm cells form muscle while others go on to make bones, cartilage or other tissues. The ultimate fate of the cells depends on where they end up in the embryo and the signals they receive from the cells around them.

Scientists are still trying to crack the underlying code that determines exactly which transcription factors bind to which sequences throughout the genome. It is not always easy to identify enhancers based on DNA sequence alone, or predict whether they will be active in a given cell type. Adding to the complication is the discovery that some enhancers are located close to the gene they act upon, while others can be a long way away – in some cases, many thousands of bases from the start of the gene. To find potential control switches, researchers tend to use techniques for capturing stretches of DNA stuck to transcription factors, or other molecules associated with active enhancers. By comparing these DNA sequences between different types of cell, scientists can begin to work out which regions are likely to be the important control switches for genes that give that cell its own identity.

Ol' Blue Eyes

Human eye colour is influenced by more than ten genes, but a single change in a control switch lies behind all the blue-eyed people in the world. The key player is a gene called "OCA2", located on chromosome 15, which encodes a protein known as P. This is one of the key factors determining how much melanin pigment a person makes in the coloured ring-shaped iris of his or her eye. Some versions of OCA2 lead to the production of more melanin while others make less. In 2008, scientists at the University of Copenhagen discovered that all blue-eyed people have the same change in a non-coding intron in another gene called "HERC2", which sits right next to OCA2. This region turned out to contain a crucial control switch that normally turns on OCA2 in the iris. In blue-eyed people, this switch is altered so they do not turn on the gene, nor do they make the P protein or produce any eye colour pigment.

Carrying two copies of this altered version is guaranteed to give someone blue eyes, although the exact shade is tempered by the mix of variations that are carried in other pigment genes. Because every single blue-eyed person studied so far carries the same genetic change, it must have come from just one original founder. We do not know who they were, but the oldest-known remains from a blue-eyed human are the fossilized bones of a man found in a cave in Spain, and dating back around 7,000 years. Analysis of his DNA revealed that he probably had blue eyes, dark brown skin and was also lactose intolerant.

When it comes to mapping out gene regulation, the biggest challenge is to obtain a pure enough sample of cells to work with. It used to be said that humans have about 200 different types of cell, going by what our tissues and organs look like under the microscope. Based on patterns of gene activity, it is likely that we actually have several thousand varieties of cell. Yet, previous studies used relatively large tissue samples, which contain several different types of cells. For example, even a relatively small chunk of brain or heart can still contain many millions of cells and several distinct cell types. So the specific patterns of gene activity produced by each individual cell type all get merged together, confusing the results. It is like looking at a large crowd of people from a distance and just seeing grey shirts. A closer look at a few of them reveals that some are wearing black shirts while others are wearing white. There are even a few wearing brown, which cannot be seen at all from far away. Advances in technology now mean that scientists can start looking for transcription patterns in very small samples – or even single cells in some cases – which will provide a lot more detail about the control switches in the genome and how they work.

LOOP THE LOOP

Almost every cell in the body contains 2.2 metres (just over seven feet) of DNA, packed into the nucleus. Rather than being neatly organized – like the neat X-shapes made by condensed chromosomes in dividing cells (see Chapter 4) – it is tangled and twisted like a bowl of biological noodles. Each one of the 46 chromosomes is squashed into its own space or chromosome territory, as it is known. Active genes are usually found closer to the centre of the nucleus than inactive ones (although this is not always strictly true). Rather than being fixed and static, DNA constantly wriggles around, along with the transcription factors, RNA polymerase and other molecules packed in there too.

All this looping and wiggling helps to explain the workings of our genetic control switches and of enhancers located a long way from the gene that they act upon. DNA can bend and twist around, bringing the start of a gene in contact with an enhancer loaded with transcription factors and RNA polymerase, so transcription can begin. However, it is not entirely clear whether the enhancer and the gene need to touch each other physically, or if they just need to be in the same general area. This also explains the fact that transcription from any given gene tends to occur in short bursts, rather than in a continuous production line. Many molecular parts – including DNA, transcription factors and all the components of RNA polymerase – need to be assembled in order for a gene to be transcribed, and they are held together by relatively weak interactions. As long as everything is in place, transcription will happen, but it does not take much for the components to fall apart and transcription to stop.

EVOLUTION'S PLAYGROUND

In 1975, the American geneticist Mary-Claire King calculated that the differences between human and chimp proteins are too small to account for those between humans and our primate cousins. She realized that it must be how these genes are controlled – not the proteins that they make – that give us our uniquely human characteristics. Scientists now know that although we separated from our primate ancestors millions of years ago, our genomes are remarkably similar, especially the protein-coding genes themselves. Nevertheless, there are crucial differences in many control switches and other non-coding sequences that affect when and where genes are switched on in humans, compared with non-human primates. These differences are what distinguish us and make us human.

In fact, control switches are evolution's playground, as Dr Duncan Odom and his team at the University of Cambridge have discovered. They studied the genomes of 20 different types of mammals, ranging from humans, mice and cows to naked mole rats and whales, which all evolved from a common ancestor over the course of 65 million years. Although the sequences of protein-coding genes tend to be very similar across all these species, Dr Odom and his team noticed that the control switches have changed a great deal during that time. Many of our genes carry

Fingers and thumbs

Anyone visiting the Florida estate of the writer Ernest Hemingway cannot fail to notice the multi-toed cats that roam the house and gardens. These so-called polydactyl cats appear to have thumbs, making their paws look like cute, furry mittens. Having examined their DNA, Professor Bob Hill and his team at the University of Edinburgh found that their extra toes are due to a fault in a control switch for a gene called Sonic Hedgehog, which helps to direct the formation of toes as a kitten grows in the womb. The change alters the timing of Sonic Hedgehog activity at a crucial point in paw development, meaning that too many toes are made. Humans also have a version of the Sonic Hedgehog gene. Hill and his team discovered that similar changes in the control switch that turns on the gene as a fetus grows in the womb leads to babies born with extra fingers and toes.

out several roles within the body, particularly during gestation in the womb. So, any major changes in the protein-coding sequence of a gene could have serious negative effects on a growing embryo. The results of this study demonstrate that the sheer diversity of mammal species in existence today are more likely to have resulted from changes to when and where genes are turned on and off than alterations in proteins. Closer to home, there are many examples of how changes in control switches have affected human traits and characteristics, including the ability to drink milk into adulthood, having blue eyes and even the colour of hair and skin (see boxes on pages 73 and 75).

So far, we have learned how control switches and transcription factors turn genes on at the right time and in the right place, or in response to changes in the environment. The next chapter will look at special molecular "sticky notes" pinned onto DNA that help to lock in patterns of gene activity, enabling cells to remember which genes are supposed to be used and which should stay silent.

S T I C K Y N O T E S

It is not enough just to switch on genes in the right place at the right time. Cells also need to remember which genes are active and which should remain quiet.

All cells contain the same set of DNA, yet they do different jobs within the body. Liver cells switch on liver-specific genes while muscle cells turn on muscle genes, and so on. These patterns of gene activity go all the way back to the very start of life, as a person grows from a single fertilized egg into a fetus, and then from a baby into an adult.

As we will discover in Chapter 11, human beings start out as a small clump of identical stem cells. However, it is not just a question of activating the right genes at the right time. It is vital that cells remember their decisions and maintain the correct pattern of gene activity. A skin cell should not suddenly switch off genes that make the sturdy proteins that hold us together and start making digestive enzymes instead.

In the same way, a nerve cell in the brain should never manufacture muscle proteins.

As we saw in the previous chapter, genes are switched on by groups of proteins called transcription factors that bind to special DNA sequences (enhancers). These act as switches, prompting the gene-reading machinery to start transcribing a gene into RNA. There is also another layer of complexity, over and above the genetic data encoded in the letters of our DNA, and this is known as epigenetics.

If the DNA sequence of our genes acts as recipes for making proteins, then epigenetic information in the genome is like a set of sticky notes or highlighter pens, reminding cells which collection of genes they should be using. There are two main types of epigenetic marks within our cells: histone modification (molecular "tags" on the proteins that package DNA) and DNA methylation (chemical modification of the DNA itself).

GET PACKING

Packed into the nucleus of every cell are more than two metres (seven feet) of DNA broken down into 23 pairs of chromosomes. In order to fit everything in and prevent it from getting tangled up, the DNA of each chromosome is wrapped around ball-shaped proteins called histones. Collectively, this makes up a chain of proteins and DNA that looks a bit like beads on a string. Referred to as a nucleosome, each "bead" is composed of four histone proteins wrapped twice round with 146 base pairs of DNA, each one sandwiched by 20 to 80 base pairs.

This mixture of DNA and histones is usually referred to as chromatin. This was first named by the nineteenth-century German biologist Walther Flemming, when he noticed thread-like structures within the nucleus that stained easily using coloured dyes. Because DNA is a double helix, it has a tendency to twist up on itself, so the strings of nucleosomes coil and stack together to form more organized structures. When a cell is ready to divide, these coils are twisted up even further, forming the X-shaped chromosomes described in Chapter 4.

As well as having a ball-shaped core, around which DNA wraps itself, histones have "tails" that protrude from the nucleosome. Certain places within these tails are prime locations for chemical "tags" known as histone modifications. Examples of common histone modifications include acetylation, methylation, phosphorylation, ubiquitination and sumoylation – each of which refers to a particular type of tag.

What's in a Word?

The term epigenetics, meaning "above genetics", was first coined by the British biologist Conrad Waddington during the mid-twentieth century. While studying fruit flies, he noticed that making environmental changes – such as altering the temperature or exposing the flies to a particular chemical – affected the appearance of the insects, suggesting that the environment could somehow communicate with the genes. Today, the word is mainly used to describe the ways in which your DNA, the proteins and RNA it encodes, and the environment within and around the body all come together to make you who you are. Some scientists feel that it should only be used to describe changes in gene activity that are passed on as cells divide, or handed down the generations from parents to children. Others use the word epigenetics as a catch-all for anything that affects how genes are switched on and off.

In turn, these different tags have specific meanings. Patterns of histone modification are thought to play an important role in how cells remember which genes should be switched on or off, helping them maintain their identity. For example, histone acetylation attracts proteins that open up chromatin, making it easier for RNA polymerase and the gene-reading machinery to access DNA, so that genes can be read. In fact, certain types of histone methylation are associated with tightly packed, closed chromatin, so genes in these regions are effectively switched off and shut down.

MARKING METHYLATION

DNA is composed of four repeated chemical bases ("letters"): adenine (A), cytosine (C), thymine (T) and guanine (G). As far back as the 1940s, researchers discovered that in addition to the regular version of cytosine, a slightly different form – referred to as 5-methyl cytosine or meC (often just called DNA methylation) – is also present and carries a little chemical tag known as a methyl group. This altered version makes up around 1 per cent of all the DNA bases within the human genome.

Importantly, these meC letters are not randomly scattered throughout our genes; they are found in very specific places. Scientists have spent a great deal of time mapping the patterns of meC in various tissues and organs, including diseases such as cancer, and comparing methylation patterns between different people. Mostly, DNA methylation tends to occur in the non-coding part of the genome, especially those sections containing repeated, long-dead, virus-like sequences (known as retroelements). Scientists think that methylating such sequences prevents them from jumping around in the genome and causing damage, or accidentally activating neighbouring genes (see Chapter 15).

There are also stretches of meC near the beginning of certain genes, particularly imprinted genes (see Chapter 14) and also those that play an important role during development in the womb. DNA methylation is often assumed to be involved in switching genes off by attracting proteins that help to shut down chromatin. Nonetheless, no correlation between silent genes and DNA methylation has yet been found, so researchers are still unclear as to how this modification affects gene activity.

Some scientists think that epigenetic marks, such as DNA methylation and histone modifications, play a significant role in directly controlling whether genes are switched on or off. Others suspect that the underlying DNA sequence and the transcription factors that bind to it are much more important in directing gene activity. They argue that epigenetic marks act more like signposts or notices, providing extra information about an underlying gene. This is a bit like a closed sign on a locked shop door – it is the lock rather than the sign itself that keeps customers out, but the notice provides extra information so customers know not to try to enter in the first place.

NATURE MEETS NURTURE

Unlike the underlying DNA code, epigenetic marks are not permanent, precise or fixed. They can easily be put on and taken off by special enzymes that respond to signals coming into cells from elsewhere. This could be the environment within the body itself – such as an infection – or changes in the outside world, such as a rise in temperature or the presence of harmful chemicals. In this way, epigenetics provides an important bridge between nature (genes) and nurture (the environment). In fact, it is often seen as the way in which the environment "talks" to our genome, enabling cells to respond to the many and varied changes and challenges that life brings about.

Understanding the epigenetic interplay between nature and nurture reveals why our genes are not the perfect code that tells cells exactly what to do. It also explains why identical twins are not exact duplicates, even though they share the same genes. Right from the very beginning of development there may be subtle epigenetic differences between them. As the twins grow and age, these differences build up, while unique genetic mutations are also acquired along the way (*see* Chapter 4).

Furthermore, epigenetic changes may be involved in many diseases, including cancer (*see* box on page 84). Across the world, researchers are busy investigating how this works and finding ways to manipulate epigenetic marks through diet or drugs (*see* box on page 85), although this is still at a relatively early stage.

DOWN THE GENERATIONS

For centuries, scientists have wondered whether characteristics acquired by people during their lifetime can be passed on down the generations. Jean-Baptiste Lamarck, the eighteenth-century French biologist, believed that this was the case. As an example, he thought that a giraffe's long neck is the result of its ancestors' habit of stretching up to high branches to reach the tastiest leaves. However, his ideas were discredited once DNA was found to be responsible for passing genetic information

Epigenetics and Cancer

Changes in genes controlling the division and death of cells are a powerful driver for cancer (*see* Chapter 16), but it is now becoming clear that tumours have altered patterns of epigenetic modification, including DNA methylation and histone modification, at important genes. Researchers are now developing and testing drugs aimed at altering or re-setting these marks. Such drugs are showing promise in treating some types of tumour (particularly blood cancers). Intriguingly, a 2015 study showed that a drug that removes DNA methylation actually works by fooling cells into thinking that they have been infected by a virus, rather than changing the methylation marks at specific genes. Of course, there is a long way to go before epigenetic therapy becomes a mainstay of cancer treatment, but it is nonetheless an exciting and growing field.

from parents to their offspring, with such characteristics being selected through evolution. There is also growing evidence that epigenetic marks of some sort may also pass from generation to generation, providing a way for environmental changes that befall parents to influence the characteristics of their children.

One of the best examples of this concerns mice that carry a gene known as "*Agouti* viable yellow", which affects coat colour. Led by eminent molecular biologist Professor Emma Whitelaw, researchers in Australia discovered a stretch of DNA close to the gene that is a good target for DNA methylation. A lack of methylation indicates that the gene is more likely to be switched on, resulting in pale or yellow-coloured fur, while increased methylation – and a darker coat – means the gene is probably switched off. Feeding pregnant female mice supplements to increase DNA methylation results in more pups being born with darker coats, while giving them a different chemical to remove methylation leads to yellower pups. Furthermore, these shifts in methylation and gene activity also seem to be passed on down the generations – a phenomenon known as transgenerational epigenetic inheritance.

Scientists are now searching for evidence that changes in a human parent's diet, lifestyle and environment can epigenetically influence gene

Pimp My Genome

Following on from the discovery that the environment talks to our genes through epigenetic mechanisms, scientists are investigating whether we can alter epigenetic marks in our cells by changing our behaviour and, in turn, improve our health. Much of this work has been done using animals, such as mice, as models, but work is increasingly being done on humans too. For example, one team of researchers in Sweden has shown that putting volunteers through a six-month exercise programme can change the patterns of DNA methylation in their fat cells. Other researchers are looking into whether chemicals in foods, such as broccoli, red grapes, dietary fibre or even garlic, can alter epigenetic modifications – a field known as nutrigenomics.

Canadian scientists have suggested that an abusive childhood can leave tell-tale epigenetic marks on hundreds of genes. Another theory, from lifestyle gurus, is that simply changing your thoughts or emotions can influence your epigenetics, although there is no convincing evidence to prove this. The big challenges here are to prove that such lifestyle changes definitely alter epigenetic marks (rather than the epigenetic changes reflecting some other process at work), to show that these changes affect gene activity, and then to reveal whether this influences long-term health and the risk of disease. There is a great deal of research to do before all these questions are answered.

activity in their children, grandchildren, or even great-grandchildren, in a similar way. Perhaps the strongest example comes from the Dutch "Hunger Winter" – a short period of severe famine in the Netherlands at the end of the Second World War. Babies born to women who were in the early stages of pregnancy during the famine have grown up to have an increased risk of certain health conditions, including obesity, diabetes and heart problems. There is also proof that their children (the grandchildren of the original mothers) have similar problems, suggesting that epigenetic changes that occurred in response to the famine may have been passed on.

Other studies have claimed that traumatic life events, such as the Holocaust during the Second World War or the 9/11 terrorist attacks in New York, can produce inherited epigenetic changes. In general, researchers are not convinced by the results, so more detailed work needs to be done to prove that this really occurs.

It is also unclear how epigenetic information is passed from parent to child. One theory is that epigenetic marks, such as DNA methylation and histone modifications, are wiped out in germ cells (the cells that make eggs and sperm). However, new results from studies of human germ cells suggest that at least some DNA methylation marks may sneak through. Some researchers think that tiny fragments of RNA may travel in eggs and sperm, affecting the activity of certain genes in response to environmental changes (see Chapter 10). There is still a lot more to learn about how transgenerational epigenetic inheritance works – if it occurs at all in humans – and this is a very exciting new area of science.

We've just taken a look at how epigenetic marks can influence gene activity. Now let's take a closer look at how RNA itself can switch genes off.

CHAPTER TEN

THE RNA

WORLD

**From fine-tuning gene activity to forming
the basis of future therapies, RNA is much
more than just a molecular messenger.**

So far we have encountered RNA – ribonucleic acid – in the form of a molecular messenger, bridging the gap between genes and the proteins they encode (*see* Chapter 3). Of course, this is an important role, but it is not the only one. Appearing in a wide variety of shapes and sizes, RNA does many different jobs inside cells. It can pair up with itself, creating double-stranded RNA or larger, more complex three-dimensional shapes, as well as combining with single strands of DNA to produce RNA/DNA hybrids. It can also team up with proteins to form biological "machines" known as ribozymes, which carry out important jobs inside cells. A good example of a ribozyme is the similar-sounding ribosome – the molecular machine responsible for building proteins. Ribozymes were first discovered by US scientists Sidney Altman and Thomas Cech, who won the 1989 Nobel Prize in Chemistry for unravelling some of the mysteries of RNA.

This all means that there is a great deal of RNA floating around in our cells. Less than 2 per cent of the human genome is composed of protein-coding genes that are transcribed into messenger RNA, yet scientists have also discovered that many other regions of DNA are transcribed into RNA. This adds up to a significant portion of the genome, but in many cases it is not clear what all this RNA is doing or why so much of it is made.

MORE THAN A MESSAGE

In Chapter 3 we saw how genes are read by RNA polymerase to produce RNA transcripts. These are spliced, capped and tailed, creating fully processed messenger RNA (mRNA) that is translated by ribosomes to make proteins. Although most genes carry instructions for making proteins, some genes are transcribed into RNA that is never translated, known as non-coding RNA. Instead, the RNA itself plays an important role within the cell. For example, ribosomes themselves contain large pieces of RNA, produced from special ribosomal RNA genes. There are also the transfer RNAs (tRNAs) that bring in amino-acid building blocks whenever proteins are made (see Chapter 3). As another example, the process of X-inactivation – where one of the two X chromosomes in female cells is permanently switched off – depends on an untranslated RNA called XIST (see Chapter 14).

There are other types of non-coding RNA too, and this is where it gets complicated. Scientists have found several different classes of RNA, ranging from very short to very long, transcribed from many regions of the genome. In some cases – for example, the little fragments of RNA called siRNA and microRNA (see below) – we know how these molecules work inside cells. However, there are also numerous long non-coding RNAs, known as lncRNAs, many of which have unknown functions. Some are transcribed from proper genes but in the opposite direction (so-called antisense RNAs), while others are read from regions of DNA in-between genes (long intergenic non-coding RNAs, or lincRNAs) or from the control switches that turn genes on and off (enhancer RNAs). A further group come from long-dead "zombie" genes, known as pseudogenes (see box on page 93).

Many thousands of lncRNAs have been found in human cells and estimates of their numbers range from around 5,000 to upwards of 50,000. In most cases, these lncRNAs make up just a very small proportion of all the RNAs inside a cell. This has led some scientists to believe that they might not be all that important. Because there are significant amounts of RNA polymerase within the nucleus and a lot of transcription occurring, it is possible that some regions of DNA are read by accident as the gene-reading machinery passes by.

Another theory is that some non-coding RNAs play an important role in organizing DNA and gene activity inside the nucleus. They could act as signposts, helping to bring in proteins that switch genes on or off. They

---

RNA before DNA?

If all the molecules within cells held a popularity contest, DNA would surely win. We hear about DNA in the media, advertising campaigns, TV shows and films. Whoever talks about a talent for singing or painting being "in my RNA"? Yet there is growing evidence that RNA was around on Earth long before DNA or proteins. More than 40 years ago, such scientists as Francis Crick (one of the co-discoverers of the structure of DNA) suggested that the complex three-dimensional structures formed by RNA could be used to carry out such biological jobs as joining molecules together.

This is supported by the fact that the main functional parts of the ribosome – the key protein-building machine within cells – are composed of RNA rather than protein. RNA can encode information in a sequence of letters and exactly replicate itself, which is the fundamental basis of life, after all. Although it is broadly accepted that RNA came before DNA and protein, it remains unclear whether or not it was the very first complex molecule to evolve. There may have been other steps along the way beforehand, based on even simpler molecules than DNA and RNA, although this idea is controversial.

---

might also work rather like the child's string game Cat's Cradle. According to this idea, one lncRNA sticks to a particular stretch of DNA, forming an attachment site for special "grabbing" proteins that can pull the DNA into the correct three-dimensional shape for genes to be activated, like fingers holding onto the strings of the cradle.

In the past, it was difficult to study RNA because this breaks down much faster than DNA. Now that more sensitive RNA-sequencing techniques have been developed it is easier to study tiny amounts of RNA within relatively small samples of cells. So the list of lncRNAs is growing all the time, even though it is still unclear whether all these newly discovered transcripts actually have a function or not. It is certainly a very hot topic of research right now, and scientists all over the world are trying to work out which of these lncRNAs are useful and which are just "junk" RNA, similar to junk DNA in the genome.

SMALL BUT POWERFUL

Several types of short non-coding RNAs exist alongside these long non-coding RNAs. Discovered in the 1980s, they play an important role in controlling how genes work. At the time, researchers in Arizona were trying to change the colour of petunia flowers by adding in extra purple pigment genes to the plants. They discovered that adding many copies of the pigment gene did not turn the flowers a deep shade of purple, but instead resulted in unusual purple-and-white patterns on the petals. So, rather than producing large amounts of pigment, the purple gene seemed to be switched *off* within some of the cells making up each flower.

Italian researchers noticed that something similar occurred when they added extra colour genes into bread mould. Instead of making the fungus a darker colour, the extra genes seemed to switch off the mould's own colour genes completely.

Pulling together all these strange observations, Andy Fire and Craig Mello (two other US scientists) carried out their own experiments with tiny nematode worms. They discovered a phenomenon known as RNA interference, or RNAi, which relies on antisense RNA, designed to pair up with the RNA transcribed from a gene. If antisense RNA against a particular gene is put into cells, it pairs up with its corresponding messenger RNA, creating double-stranded RNA. This type of RNA is usually only found in viruses, so cells will try to slice it up as if they were protecting themselves from viral infection. However, in doing so, they actually cut up the messenger RNA for the gene (as well as the antisense RNA) so this cannot be translated into protein. In the case of the purple petunias, the white patches on the petals are caused by a lack of pigment protein, because all the pigment gene messenger RNA has been destroyed.

There is another step that makes RNA interference even more effective. These small fragments of sliced-up double-stranded RNA, known as siRNA, can also work together to silence proteins inside the nucleus and shut down transcription altogether. It stops production of a protein from mRNA and also prevents the gene from being read. Further experiments revealed that just adding the right siRNA fragments directly to cells – or even feeding it to worms – is enough to switch off any gene. It does not just apply to worms either; RNA interference works within all animals, including human cells.

Fire and Mello won the 2006 Nobel Prize in Physiology or Medicine for their discovery of RNA interference, which has turned out to be an

Going Round in Circles

So far, we have encountered RNA as straight lines of letters – long strings of messenger RNA or long non-coding RNA, and also as short fragments of siRNA and microRNA. However, yet another form of RNA was discovered much more recently. In 2012, Dr Julia Salzman and her team at Stanford University in California published a study showing that hundreds of human genes produce RNA that is formed into perfect circles. The origin and function of these is still not known, although some scientists believe them to be a kind of molecular sponge for mopping up excess microRNAs. Despite the fact that circular RNA is produced in large amounts, nobody had noticed it before and so its discovery was a big surprise. Maybe there is yet another type of RNA lurking undiscovered within our cells?

incredibly useful tool for biologists. Each siRNA is just 21 letters long and can be made in a test tube using chemical reactions. In theory, therefore, it is possible to switch off (knock down) any gene in any cell type. This technique has made it quick and easy for scientists to knock down genes to see what they do, either in cells growing in the laboratory or within such small organisms as worms or fruit flies. Because of its impressive power, pharmaceutical companies have been very interested in developing RNA interference-based therapies for a wide range of diseases.

The first of these – fomivirsen – was developed to treat viral infections in people with AIDS, and was approved by the US Food & Drug Administration (FDA) in 1998. Another RNA-based drug, mipomersen, is used to treat people with hereditary high cholesterol. Nonetheless, there have been a great many failures along the way, as it is difficult to deliver double-stranded RNA to the right place in the body and then insert it into cells. Despite these disappointments, researchers are working on new delivery systems, with many new RNA-based treatments and vaccines currently undergoing clinical trials.

FINE-TUNING THE GENOME

As well as siRNAs, researchers have discovered other short double-stranded fragments of RNA, known as microRNAs. These are very slightly longer than siRNAs – 22 letters rather than 21 – and they are made in a different way. MicroRNAs are produced from longer RNAs that are transcribed from hundreds of special genes. Each one of these long transcripts is then folded up on itself to create double-stranded DNA, which is sliced up to make up to six different microRNAs. Overall, more than 1,000 microRNAs are produced from the human genome, with many being made in the brain. So, what do they actually do?

Unlike siRNAs, which must pair up exactly with the sequence of the gene they are targeting, microRNAs are less precisely matched. As a result, there are several different messenger RNAs that they could pair with, and it is possible that around 60 per cent of our genes are targeted by microRNAs. In the same way as regular RNA interference, a particular microRNA can lead to any matching messenger RNA being carved up and unable to be translated to make proteins. Other, subtler effects of microRNAs on gene-activity levels include changing epigenetic marks (*see* Chapter 9) so genes are switched off. Conversely, they can sometimes even have the opposite effect and turn genes on.

Zombie Genes

The human genome is scattered with so-called pseudogenes. These are "dead" genes that originally started out as a copy of a functional protein-coding gene, but were somehow damaged or broken at some point during our evolutionary history. Just as zombies are dead people that come back to life in horror films, some of these dead genes can also start working again and become transcribed into RNA. Although they no longer make a protein, the RNA from these zombie genes can interfere with "living" genes, altering their activity patterns and affecting the functioning of a cell. They are not quite as exciting as the characters in a scary movie, but these zombie genes could still turn out to be biological blockbusters.

Scientists currently think that microRNAs act as fine-tuners of gene activity, helping to keep everything working at the right level. It is therefore not surprising that problems with certain microRNAs have been linked to illnesses including cancer or heart disease, and such conditions as obesity and alcohol addiction. Despite their small size, microRNAs certainly make a big impact, but there is still much to learn about how these little molecules help to control genes and health.

Over the past few chapters we have looked in detail at how genes are switched on and off at the right time and in the right place. Now it is time to zoom out and see how this works during the complicated process of building a baby.

CHAPTER ELEVEN

BUILDING

A BABY

The route from a single cell to a fully grown baby is long and complicated. Happily, researchers are beginning to understand some of the genetic rules and patterns that guide us through this incredible journey.

Every human starts out as a single cell (zygote) that is created when egg and sperm meet, setting in motion a complex programme of gene activity, cell division and specialization. One cell divides into two, then four, eight, sixteen and so on, with groups of cells gradually organizing themselves into tissues and organs with different roles. Roughly 40 weeks later a baby is ready to be born, equipped with all the necessary organs for life – brain, heart, liver, lungs, kidneys and much more. Along the way, cascades of genes are switched on and off, each setting the scene for the next stage of development. It is rather like a computer game in which each level needs to be played in order to reach the next one.

Understanding how the genetic code written in our DNA translates into something as wonderful as a human being – together with the influence of the environment and epigenetics (*see* Chapter 9) – is a major challenge for developmental biologists. As well as discovering how genes direct embryonic development when all is working well, it helps to shed light on what happens when things go wrong. Up to a fifth of all known pregnancies end in miscarriage, and most of these occur in the very early stages. Although every parent hopes that their child will be born healthy, inherited or spontaneous genetic changes can lead to abnormalities and problems that range from mild to extremely severe.

For obvious ethical, moral and practical reasons, it is difficult to study the genes involved in human development. Acquiring human embryos and fetuses for research purposes is difficult, particularly in the earliest stages of pregnancy, as this is tightly regulated in many countries. Instead, scientists study smaller animals such as mice and monkeys; like us, they are mammals and develop in a very similar way. We are now going to wind the clock back to the very earliest stages of life, and look at some examples of the processes at work.

GETTING STARTED

Assuming that conception occurs in the usual way – through sexual intercourse, rather than using a technique such as *in vitro* fertilization (IVF) – the journey from egg to baby begins in one of the two fallopian tubes that connect a woman's ovaries to her womb. The embryo takes several days to travel to the womb, dividing from one cell into a tiny hollow ball made up of a few hundred cells (the blastocyst) along the way. Nestled inside is a clump of embryonic stem cells, known as the inner cell mass. Embryonic stem cells have the capacity to become any type of cell in the body, growing and specializing to form all the organs and tissues of the fetus – a characteristic referred to as pluripotency.

The next step is implantation. Once the blastocyst has made it to the womb, the cells on the outside (called the trophoblast) burrow into the wall of the womb and set about making the placenta, from where the growing fetus gets oxygen and nutrients. The stem cells keep dividing as part of their journey to become a more structured embryo. In fact, this tiny embryo is already becoming organized and specific patterns of genes are being switched on. For example, a gene called "Oct4" is important for making sure that embryonic stem cells maintain their pluripotency (the ability to develop into any type of cell or tissue), so it is only active within the inner cell mass and is switched off in the trophoblast.

UP AND DOWN, INSIDE AND OUT

Like all mammals, humans have a top and bottom, front and back, left and right (heart, stomach and spleen on the left, liver on the right). These positions are determined very early on in development. Egg cells in animals such as fish, frogs or birds have a big yolky blob in their bottom half, and this determines which way is up or down once the egg is fertilized. Human egg cells are symmetrical, so something has to happen to break the symmetry and distinguish top from bottom. Based on experiments with mice and monkeys, the place in which the sperm enters the egg seems to help determine the line along which the fertilized zygote divides into two cells. This then has a knock-on effect through further cell divisions, setting up the position of the inner cell mass in the blastocyst and, ultimately, distinguishing top from bottom.

Also established in the early embryo are three distinct layers that make up the body: ectoderm (which mainly forms our skin and hair, and also makes the nerves and brain); endoderm (which becomes guts, lungs and other tube-shaped material); and mesoderm (which makes muscle, blood and bone). These can be roughly thought of as outside, inside and middle matter. The three layers are set up around the same time as gastrulation, which occurs around 15 days into development, when the embryo is a disc-shaped collection of cells.

Gastrulation is probably the most fundamental part of early development, as it sets the stage for everything that follows. For this reason, the British biologist Lewis Wolpert once said, "It is not birth, marriage, or death, but gastrulation which is truly the most important time in your life." Because of the ethical and technical complexities involved in obtaining embryos at this stage, it is extremely difficult to study gastrulation in humans, so our best guess about the process comes as a result of studying mice.

Devastating Damage

During the 1950s, doctors began prescribing a new drug called thalidomide to pregnant women to help relieve morning sickness. It was thought to be safe, but it had never been tested properly on pregnant animals or humans. Thalidomide turned out to have devastating effects on limb development, and babies were born with deformed arms and legs. Between 5,000 and 10,000 children were affected worldwide, and more than 2,000 are still alive today. Instead of affecting the genes involved in building limbs, thalidomide interferes with the interaction of three molecules that are important for directing the formation of blood vessels in developing arms and legs. By shutting off the blood supply at a crucial point in development, thalidomide stops the limbs from growing to the right size.

Beginning as a thin stripe, called the primitive streak, running along the top of the embryo – which also distinguishes the back of the embryo from the front – cells start to move around within the embryo and sort themselves into three distinct layers. Eventually, these layers tuck and fold themselves into an organized tube-like structure, with endoderm in the centre, ectoderm on the outside and mesoderm in the middle. The cells in each of the layers then switch on specific patterns of genes according to their location and the signals that they receive from the cells around them, directing them towards certain fates while restricting other options. For example, some cells in the ectoderm set out on a journey towards becoming brain cells, switching on nerve genes and permanently shutting down genes that make muscle or gut. We will cover this process of brain development – known as neurulation – in more detail in the next chapter.

LEFT OR RIGHT?

Another important decision made by the embryo is to tell left from right. This occurs around the time of gastrulation, long before any organs have started to form. Again, our best guess as to how this happens comes from studying the process in animals such as chickens and mice, but it seems to start from a structure on top of the embryo called the node. This is covered with tiny hair-like structures called cilia, which are tilted in such a way that they waft fluid from right to left across the embryo. This current causes a signalling molecule called "Nodal" to build up on the left-hand side. In turn, Nodal switches on a set of genes – including one called "Lefty" – that tell the left side of the embryo to develop slightly differently from the right. In contrast, the absence of Nodal on the right-hand side leads to a different set of genes being activated.

Problems with the cilia, Nodal, Lefty or one of the other left-leaning genes are very rare. In this case, the organs grow in a mirror image of the rest of the population – a condition known as *situs inversus*. Famously, Doctor Julius No – the villain from the James Bond story *Dr No* – was affected in this way, and survived an attempted stabbing because his heart was on the right rather than the left.

SETTING UP SEGMENTS

Once the developing embryo has established a basic pattern – up, down, front, back, left, right and the three layers of ectoderm, mesoderm and endoderm – the next step is to divide it up into segments from top to bottom, each with its own identity. Human beings are clearly separated into sections: head, neck, torso and limbs. The spine comprises many repeating segments of bone, flanked by a regular pattern of ribs. This segmented pattern is common to all complex animals, from insects upwards, and is directed by a special group of genes called "Hox" genes.

Hox genes were first discovered in fruit flies in the early twentieth century, when American geneticist Thomas Hunt Morgan noticed that certain mutations made body parts grow in the wrong place. Some flies grew legs

For Your Eyes Only

It may not seem as if human eyes have much in common with those of a mouse, fish or fruit fly, but they do. The first step towards building an eye in all of these organisms starts with the same gene: Pax6. It acts as a sort of master controller, switching on a number of genes in the right place to begin building the structures that make up the eye. Pax6 is known as "eyeless" in fruit flies – because flies lacking the gene have no eyes at all – and humans with a faulty version of the gene suffer from a condition called aniridia, in which the iris of the eye fails to develop. The sequence of the gene is extremely similar across the animal kingdom. Amazingly, activating the mouse version of Pax6 in a fruit fly is enough to make it start growing normal fly eyes!

where their antennae should have been, while others had an extra middle section (thorax), complete with an extra set of wings. Later, scientists found Hox genes in all kinds of organisms, from jellyfish to humans, tracking them down relatively easily because their DNA sequence is very similar across the animal kingdom. However, the numbers of Hox genes differ, depending on the species. Fruit flies have just eight while humans have 39, clustered into four groups on different chromosomes.

Intriguingly, the order of the Hox genes in each cluster mirrors the pattern in the body in which they are active, from the head down to the toes. It also matches the timing of their activation, with head genes being switched on first and limb genes coming on last. They can be thought of as genetic coordinates, telling different sections of the body to adopt a particular fate – for example, to become skull, or spine, or fingers and toes. Because Hox genes are so important, they need to be very tightly controlled. As Morgan noticed with his fruit flies, any changes in the pattern of when and where they are switched on can have a major impact on the body plan. Faults in Hox genes are particularly associated with abnormal bone structures, especially in the limbs and spine.

There is not enough space in this book to go into all the details of human development and the genes that direct it. Now that we have covered the basics of the body plan, we will look in more detail at building the most important organ of all: the brain.

CHAPTER TWELVE

WIRING   THE   BRAIN

The human brain is probably the most complex object in the universe. Yet, although it is constantly shaped by our experiences and environment, genes play an important role in building this powerful biological computer.

It is incredible to think that every human who ever lived grew from a tiny cluster of embryonic stem cells, as we discovered in the previous chapter. Even more amazing is the idea that the brain - with all its knowledge and complexity - originated from a little collection of cells in the early embryo. An adult brain contains around 100 billion nerve cells (neurons), each wrapped up with insulating cells and supporting cells. A neuron can make more than 1,000 connections (synapses) with other cells, adding up to around 60 trillion in total, all buzzing with messages and information. Far from being a messy tangle of interconnected nerve cells, the brain is highly organized and comprises distinct regions.

At the front is the cerebral cortex or forebrain, made of layered sheets of nerve cells. These are folded up to squash as many cells as possible into the space available in the skull, creating the characteristic ridged appearance that you see in pictures of the brain. This is the main area responsible for conscious thought and behaviour. In the middle is the midbrain, which helps to co-ordinate the body's responses to information coming in from the world around us. At the back is the hindbrain, which contains the cerebellum – which is essential for movement – and structures that control breathing, heart rate and other basic functions that keep a person alive.

BUILDING A BRAIN

The brain begins to form around about the time of gastrulation (*see* Chapter 11), when the embryo folds up to make three layers of different cells – ectoderm (outside), mesoderm (middle) and endoderm (inside). This process begins from a stripe of cells called the primitive streak, which also marks the place where the structures that will eventually form the spinal cord and brain first begin to grow. The first step in the brain-building process – known as neurulation – is the formation of a groove in the sheet of ectoderm cells along the top of the embryo, following the primitive streak. This becomes deeper and deeper, eventually zipping up to form a hollow neural tube running from top to bottom inside what will become the back of the fetus.

As the embryo continues to develop, the head end of the neural tube swells up into three fluid-filled bulges, which will eventually become the forebrain, midbrain and hindbrain. The rest of the tube forms the spinal cord – the main route through which all the nerves in the body wire into the brain. This is all directed by many different genes and proteins that work together, sending signals between cells to make sure they move at the right time and into the right place. When things go wrong with such a fundamental process the consequences can be very serious.

If the structures at the head end do not form correctly, the fetus will develop without a proper brain – a fatal condition called anencephaly.

Another potential problem is spina bifida, which occurs when the bottom end of the neural tube fails to close properly. This is a less serious, although still challenging, disability and many people with spina bifida can live into adulthood if they have the right treatment and support. Neural tube problems are linked with low levels of the vitamin folic acid (folate), so women are now advised to take folic-acid supplements if they are pregnant or trying for a baby.

The next step is to fill out the three bulges with neurons and organize them into the correct places in the brain. Within the neural tube are special stem cells called neural progenitors, which are the source of all the different types of nerve cells in the brain and body. These stem cells divide many times, creating millions and then billions of neurons that switch on specific patterns of genes, according to the type of nerve cell they need to become. Some end up in the various structures of the brain, while others

Bigger Equals Better?

Human brains are larger than might be expected for our size, compared to other primates. For example, our brains contain roughly 86 billion nerve cells – that is roughly 12 times the number within the brain of a monkey such as a macaque, yet we are only eight or so times bigger. For a long time, this was used to explain our superior intelligence in comparison with other animals. However, mice and birds have similar ratios of brain to body size as humans, yet they are not as smart. This is because human brains are built differently to those of other animals. We have a much larger cerebral cortex – the section at the front used for thought and language – as well as other unique adaptations.

Our Neanderthal ancestors actually had bigger brains than modern humans. These measured an impressive 1,600 cubic centimetres, compared with an average of 1,440 cubic centimetres for modern men, and around 1,330 for women. Yet the evidence suggests that one of the reasons that the Neanderthals died out is because humans outsmarted them, so their bigger brains did not result in a greater capacity for survival. Having bigger brains does not mean that men are cleverer than women – research shows that there is little correlation between brain size and intelligence between the sexes across the general population.

The Language Gene

During the 1980s, a family known by the initials KE came to the attention of researchers. Many family members across three generations suffered with speech and language problems, having inherited the condition in a way that strongly pointed to a single faulty gene. Eventually, scientists at the University of Oxford tracked down the culprit: "Forkhead box protein P2", or "FOXP2" as it is more commonly known. The FOXP2 gene encodes a transcription factor that turns on genes in the brain, heart, stomach and lungs in the developing fetus and later on into childhood and adulthood. Within the brain, this gene seems to be involved in wiring up regions of the brain implicated in language, perhaps helping us to master the complex thoughts and coordinated movements needed for speech.

Intriguingly, the gene is found in many other animals, and is thought to be important for vocal sounds in mice, bats, birds and more. The chimp "FoxP" protein differs from the human version by just two amino acids, but even this small difference might be part of the explanation for why humans can speak while chimps cannot. When researchers put either the human or chimp version of the gene into human brain cells grown in the laboratory, they found that it switched on different patterns of genes. FOXP2 probably acts as some kind of master controller, activating a pattern of gene activity that eventually leads to human brains being wired for speech, while chimp brains are connected up in a different way.

grow long tails (called axons) that grow out into the body, ready to receive signals from the senses or send out instructions to move muscles.

In human embryos, these neural progenitor cells can divide multiple times, creating many nerve cells, but in mice they only divide once. This means that mice have far fewer brain cells than we do, particularly in a region called the neocortex, which lies under the surface of the brain and plays an important role in conscious thought, reasoning and language. As a result, humans have much more folding in the brain than mice, and so we are more intelligent. In 2015, German researchers discovered that a small gene called "ARHGAP11B" was very active in human neural progenitors, but was not present at all in mice. When the scientists added this human gene into mice brains using genetic engineering, the animals grew many more neurons in the neocortex, and it started to fold up in the same way as a human brain – although they did not check whether the mice were any cleverer.

All the basic structures of the brain are in place by the eighth week of pregnancy, and then it is just a matter of making more and more neurons and connecting them all up. In fact, the production of new nerve cells in the brain carries on after birth, and even into adulthood. Contrary to the popular belief that grown-ups do not make new neurons, some are still made in certain places in the adult brain – notably the hippocampus, which is a key player in learning and memory. However, it is not enough to have huge numbers of neurons and a nice big brain (*see* box on page 102) to make humans special. What is really important is to understand how all these nerve cells are wired together.

WIRING IT UP

The connection between two nerve cells is called a synapse, and a single neuron can make hundreds or even thousands of connections with the cells around it. The more synapses a neuron makes, the more information it can process, and human nerve cells seem to make an exceptionally high number of connections compared to other animals. This is thought to be the reason for our exceptional brainpower, and the explanation lies firmly in the genes.

In 2012, two groups of scientists published a pair of exciting studies about "SRGAP2", a gene that makes a protein that sends signals within nerve cells and makes them grow new synapses. One team discovered that there are four copies of the gene in humans – labelled A to D – compared with just one in all other animals, including chimps and other non-human primates. These four copies arose a long way back in our evolutionary history when the original gene was somehow duplicated in the genome. This first occurred around 3.4 million years ago, making "SRGAP2A" and "B", then version "C" around 2.4 million years ago, with "D" following around a million years ago.

Left Brain, Right Brain

According to quizzes on social media, "left brain" people are organized and systematic, while "right brainers" are intuitive and creative. Although these games may be fun to do and share with friends, the concept of left brain versus right brain is not supported by science. The brain is divided into two halves (hemispheres), and imaging studies have shown that specific tasks are divided across them in three distinct patterns. Some tasks, such as general attention, are split equally between both halves, while movement is controlled by opposite hemispheres to the body (so moving your left side is controlled by the right side of your brain, and vice versa). Other aspects do seem to be focused in one half or the other. For example, spatial processing – the ability to figure out shapes and spaces – is mostly in the right hemisphere, and so are responses to music.

Math and language skills are mainly focused in the left, reinforcing the left brain/right brain idea. However, this is an oversimplification of the interconnected way that both halves of the brain work together, with information flowing between them via a thick band of nerves called the corpus callosum. In addition, brain scan studies of over a thousand people have shown that although some tasks tend to be done on the left or the right side, there is no overall bias to one particular hemisphere. So you can use both sides of your brain to realize that this idea of being a "lefty" or a "righty" is little more than a myth.

The second team of scientists delved into the function of the four different genes. They showed that SRGAP2A and SRGAP2C are very active within nerve cells, while the other two are much quieter and thereby likely to be less important. They also discovered that SRGAP2A helps to speed up the formation of synapses, but also limits the number of connections that an individual neuron can form. However, SRGAP2C counteracts this, allowing cells to develop more synapses and slowing down their formation and maturation. Together, the two genes find a balance that enables human neurons to form numerous strong synapses with their neighbours, creating the dense network of connections that gives us our brainpower.

Neuroscientists are now using the very latest brain scanning and imaging techniques to draw up detailed maps of all the connections between the billions of cells in the brain, creating beautiful and delicate wiring diagrams known as connectomes.

NATURE OR NURTURE?

We often hear the phrase "it's in the genes" used about all kinds of personality traits – such as intelligence, aggression or an optimistic outlook on life – and conditions ranging from addiction and schizophrenia to depression or autism. What is clear from all the evidence gathered so far from genetic studies and detailed neuroscience experiments is that variations in our genes have at least some impact on the brain, behaviour and psychiatric conditions.

For example, many psychological traits and conditions are much more similar or common in identical twins (who have identical genes) than in non-identical twins, whose genes differ yet they still share the same upbringing. By mapping information from brain scans onto genetic studies, scientists are starting to unpick how variations in genes affect the size and shape of certain brain regions and structures in each person, which may begin to explain some of these differences in traits. Although genes play an important role in directing the formation of the brain – particularly early in development – they certainly do not tell the whole story.

The brain is an incredibly adaptable organ. Neuroscientists talk about it in relation to plasticity – meaning that the brain can make and remodel billions of synapses in response to experiences and information gained throughout life. This is particularly important in early childhood when millions of neurons are still being produced and billions of connections are wiring up. So, although genes (nature) can influence the basic machinery and set-up of the biological computers within our skulls, our environment and upbringing (nurture) also adds a huge amount. Intelligence, as measured by IQ, is around 70 per cent "in the genes", based on studies comparing identical and non-identical twins. The rest depends on a huge number of factors in the environment, which range from being read to early on in life to the quality of teaching at school.

These factors have a massive influence, although each individual's genes help to set the broad range of IQ scores that they might possibly achieve. So, even if someone has inherited very smart gene variations, they are less likely to achieve their full potential without a good environment. Equally, some genetic changes and variations have a significant impact on cognitive function and learning. A great learning environment and extra help from a young age can make a big difference to the final outcome. It is also worth noting that most of us fall broadly in the middle – it is not called average intelligence for nothing!

As another example, researchers have found that people carrying a certain genetic variation on a gene called "SKA2", along with particular epigenetic marks there (see Chapter 9), are significantly more likely to suffer from post-traumatic stress disorder (PTSD) or commit suicide following childhood abuse. This outcome is by no means a certainty – it just increases the chances that someone will react badly to a terrible situation, and people who do not have abusive childhoods yet still carry the genetic variation might never experience these feelings. Similarly, people who do not carry the variation yet suffer child abuse might still go on to experience PTSD or commit suicide. We are who we are because of a combination of nature *and* nurture, and disruption of either can cause fundamental changes to our brains, bodies and behaviour.

There is still a great deal to discover about how our brains are built and wired up, as well as the way in which experiences and the environment shape how they work and make us who we are. Large-scale projects are underway to map all the cells, connections and patterns of gene activity within the brain. This is a huge task, given the sheer number of connections and cells involved, but many researchers believe that this approach is the best way to help us understand the workings of the brain.

So far we have looked at the role of genes in building the body and brain. Next, we will look at another system where genes are very important – in fact, they can make the difference between life and death.

CHAPTER THIRTEEN

COMPATIBILITY

GENES

Our genes do not just shape our bodies and brains – they also govern our compatibility with the people around us.

In 1628, the English doctor William Harvey worked out how blood circulates around the body. Shortly afterwards, people started experimenting with blood transfusions, mainly with dogs, transferring the vital red fluid from one animal to another. It was reasonably successful, but tests involving lambs' blood being put into humans were banned by the late 1600s because of serious adverse reactions.

Over the next 200 years or so, doctors developed cleaner techniques that reduced the chances of infection (even experimenting with milk as a suitable replacement for blood, which it was not). Sometimes the transplants saved lives, but often the recipient would react unfavourably to the donor's blood and many people died. The breakthrough in understanding why this happened came in 1900, when an Austrian doctor called Karl Landsteiner realized that mixing blood from some pairs of people would result in a clumping together of the cells (the sign of a reaction), while other pairs were not affected in this way. He worked out that there are three different types of blood, calling them A, B and C (now known as O). A further type – AB – was found a couple of years later. Landsteiner won the 1930 Nobel Prize for Physiology or Medicine for his discovery, which sparked a major change in our understanding of blood transfusions and saved countless lives.

For centuries, and with very little success, doctors have also been trying to transplant whole body parts. It was well known that a skin graft taken from one place could grow elsewhere on a person's body, but skin from someone else would be rejected and die. Organ transplants were even more challenging and only worked between identical twins. Clearly, there are certain factors within the body that determine whether someone's blood or organs might match those of a potential recipient. As one might expect, these are encoded within our genes.

IN THE BLOOD

Although there are several medically useful blood components, such as plasma (the fluid in which blood cells are suspended) and blood-clotting platelets, blood transfusions use oxygen-carrying red cells. These cells are coated with sugary molecules, known as antigens, which protrude from their surface. Antigens come in two forms – A and B. People with blood group A only have A antigens on their red cells, while those who are group B have just B antigens. Possessing both makes a person AB, while owning neither means they are blood group O.

The A and B antigens are made by two slightly different versions of an enzyme called a glycosyltransferase, encoded by a gene called "ABO". One version (allele) of ABO makes the glycosyltransferase that produces A antigens, another makes the glycosyltransferase responsible for B antigens, while a third allele (O) makes a glycosyltransferase that does not work at all so it cannot make any antigens. Because we inherit two copies of ABO – one each from both mother and father – there are several different combinations of these three alleles. People who are blood group

A will have AA or AO allele pairings, while blood group B people could be BB or BO. Blood group AB can only come from having one of each allele, while those who are blood group O must have two O versions.

Although red blood cells do not contain any DNA, they are still packed with proteins, including glycosyltransferases that are made as they mature.

As well as the ABO antigen system, another molecule must be taken into account. Rhesus factor is a protein found on the surface of red blood cells, and although there are around 50 different Rhesus antigens, the most important is the D antigen. It is encoded by one gene, "RHD", which either makes the protein or does not, depending on which version has been inherited. Like the ABO system, having one or two functional Rhesus alleles will make someone Rhesus positive (+), while Rhesus negative (–) people must inherit two non-functional versions.

When it comes to matching blood for a transfusion, both the blood group and the Rhesus type must be taken into account. The immune system is primed to recognize aspects that do not look exactly like us, so putting blood cells coated with A antigens into someone who is AB, B or O will trigger an immune response, as will putting Rhesus antigens into a Rhesus negative recipient. This reaction causes the blood cells to clump together. In severe cases this can be fatal, and explains why the early attempts at unmatched blood transfusions went so badly wrong. The same reaction occurs if B-type blood ends up within an A, AB or O body, or if AB blood is given to someone who is A, B or O.

There are a few exceptions to the rules of matching. Someone whose blood group is AB and who is Rhesus positive is a universal recipient. They have all the possible antigens on their blood cells, so they will not react to blood that is A, B or O, and it does not matter whether this is Rhesus positive or negative. Similarly, people who are O Rhesus negative are highly prized by blood-transfusion services for being universal donors. Their blood cells are free of any potentially triggering antigens, so they can be given to anyone who needs them in an emergency situation.

As well as playing a vital role in blood matching, the Rhesus system is also very important in pregnancy. If a woman is Rhesus negative and the father of her child is Rhesus positive, there is a chance that the

Shirley and Anthony

In 1971, Anthony Nolan was born with a rare genetic blood disorder called Wiskott–Aldrich syndrome. A bone-marrow transplant was the only hope for a cure, but there was no match within Anthony's family. Desperate to save her child's life, his mother Shirley set up a database of potential bone-marrow donors – the first of its kind in the world. Sadly, this did not help Anthony and he died at the age of eight. However, the Anthony Nolan Register now contains more than half a million potential donors, all of whom have had their compatibility genes analyzed in the hope of making a match for a desperate patient suffering from leukaemia (blood cancer) or other blood disorder. Inspired by Shirley's lead, there are now more than 19 million potential donors listed on registers in 48 countries, and more than a million bone-marrow or blood stem-cell transplants have taken place across the world.

baby could be Rhesus positive too. Usually, the foetal blood supply is kept separate from the mother's blood, but knocks and bumps can cause injuries that lead to blood mixing. If this happens, the mother's immune system starts to recognize and attack her baby's "foreign" Rhesus-positive blood cells, especially around the time of birth. This can cause anaemia and jaundice, or even lead to a miscarriage. There is also a much greater chance that the mother's immune system will harm any future pregnancies if they are also Rhesus positive. Rhesus negative women are now offered regular injections of anti-D immunoglobulin, which prevents their immune cells from recognizing and reacting to the Rhesus antigen. Since it was developed during the 1960s, this treatment has saved the lives of millions of babies.

Although matching people using the ABO Rhesus system is enough to ensure safe donations in most circumstances, blood groups are a little more complicated. There are more than 20 different sets of antigens on the surface of blood cells, all encoded by various genes and alleles. Despite this, red blood cells are still relatively simple, with a limited repertoire of molecules on the surface. It all becomes even more complex when matching organs or stem cells for transplantation.

MAKING A MATCH

Every day, immune cells are patrolling round the body checking that all is well.

They ensure that cells are healthy and functioning normally, spot and destroy invading foreigners (such as bacteria and viruses), and discard any damaged or faulty cells. To help them with this task, cells are coated with cup-shaped proteins, known as the major histocompatibility complex, or MHC. These MHC proteins take samples of molecules inside the cell and present them on the outside for immune cells to inspect – a process known as antigen presentation. If this molecular sample seems abnormal in any way, the cell will be destroyed in a process designed as a quality-control system for the body.

MHC proteins are encoded by a number of different compatibility genes, collectively known as the "human leukocyte antigen" (HLA) gene complex. These are the most diverse genes within the human genome, and there are thousands of different versions (alleles) of each one. These genes determine whether someone will be a suitable donor match for a patient needing an organ, bone marrow or blood stem-cell transplant.

Every year, thousands of people with blood cancer (leukaemia) or other blood disorders need a stem-cell transplant. This used to be done using stem-cell-rich bone marrow from a large bone such as the thigh, but 90 per cent of transplants now come from stem cells collected from the donor's bloodstream – a much less painful and invasive process. Around a third of patients can find a donor match from someone in their family, such as a brother or sister, as they are more likely to have inherited the same alleles of the HLA compatibility genes. However, because alleles are shuffled and switched when eggs and sperm are made, this is not always possible, so an unrelated donor needs to be found.

Since the 1970s, huge efforts have gone into finding potential stem-cell donors and cataloguing their compatibility genes (see box on page 111). Nowadays, transplant doctors aim to find a donor who is a ten-out-of ten match for their patients, meaning that they have the same versions of both copies of five particular HLA genes. Intriguingly, transplants can still work from donors who are not perfect matches, yet may fail even if there is a ten-out-of-ten match. This is because there are other compatibility genes at work and doctors do not currently look at these.

Restricting the search to just ten spots in the genome is rather like two potential daters filling out a simple ten-question personality quiz. Two people with ten matching answers might get on well, but there could be other aspects of each other that they do not like, and which did not feature in the quiz. Conversely, a couple with only six or seven matching answers might hit it off because they have other shared interests. Scientists are now using more detailed genetic analysis to look in depth at the five main compatibility genes, as well as investigating whether including other HLA genes could provide more accurate matching between donors and patients.

For whole organ donations, such as kidney or liver transplants, the level of HLA matching is slightly less strict. Doctors will usually look at both copies of just three HLA genes, as well as matching the ABO blood types of the patient and the donor. Matching the blood type does not matter for a blood stem-cell transplant because the transplant effectively replaces the patient's entire blood system. In fact, they could even end up changing blood type as a result of the transplant. Patients

Chasing the Immune Connection

The British scientist Peter Medawar is considered by many to be the father of transplantation. He won a share of the 1960 Nobel Prize in Physiology or Medicine for his discovery that immune cells are responsible for transplants being rejected. In 1940 he saw a bomber plane crash near his house, badly burning the pilot. He wondered why the body often rejected skin grafts for serious burns. At the time, it was blamed on the surgeons' lack of skill, but Medawar felt there must be something else. After looking at skin grafts in human burn victims and also in laboratory rabbits, he discovered that rejection occurred because the recipient's immune cells invaded and destroyed the donor tissue. This finding triggered a major research effort to understand how the immune system recognizes our own cells, as well as those of others, and led to the use of life-saving, immune-suppressing drugs and better approaches to transplant matching.

Love Stinks

The so-called "Sweaty T-Shirt Study" has become the stuff of scientific legend. During the 1990s, Swiss biological researcher Claus Wedekind asked 44 men to wear new T-shirts for two days, without using any kind of scented toiletries or deodorants. Afterwards, he put each shirt in a separate box and asked a similar number of women to sniff seven of them, rating each one as to whether they liked the smell or thought it was sexy. Importantly, Wedekind also knew each volunteer's makeup of MHC compatibility genes. Widely reported in the media, the results were striking – women preferred the smell of men whose MHC genes were different from their own.

This finding has since been corroborated by other studies, and more detailed genetic analysis suggests that people do seem to pick partners with different MHC genes, although this only seems to hold true for those with European ancestry. It is not entirely clear why this should be the case, but seeking out mates with different genes from our own might be a way to increase the chance of having a healthy baby. After all, evidence shows that having compatibility genes that are too similar can increase the chances of fertility problems or even lead to miscarriage.

are also given drugs that suppress the immune system and help to prevent transplant rejection.

Countless thousands of organ and stem-cell transplant recipients are alive today, thanks to advances in compatibility matching and immune-suppressing treatment. Despite this, there is still a desperate shortage of donors, and it can be a challenge to find matches for people with more unusual or mixed ethnic backgrounds. This problem may be solved in the future by advances in genetic engineering – particularly for donor organs (see Chapter 17) – but, for now, it is important to encourage as many people as possible to sign up as potential organ and stem-cell donors.

Our compatibility genes create variations between people that we can't see. But, as we'll see in the next chapter, the X and Y sex chromosomes are responsible for some much more visible differences.

CHAPTER FOURTEEN

X   A N D   Y

**From spotty cats to surnames, there is
more to our sex chromosomes than
meets the eye.**

The business of sex and gender is a complicated one. In the very simplest genetic and biological sense, female humans have two sex-determining X chromosomes in addition to their 22 pairs of regular chromosomes (autosomes), while males have one X and a Y. This is not an absolute rule, though. Around one in every 400 people within the general population has missing or extra sex chromosomes. Some, for example, have three X chromosomes, comprising two or more Xs and a Y or an X and two Ys, while others have just a solitary X.

Sex determination starts very early, during embryonic development in the womb. Usually, the absence of a Y chromosome results in a female baby, while the presence of the Y chromosome – or even just certain genes found on the Y – starts off the development of testes (the male sex organs). Again, this is not a 100 per cent certainty, and genetic or hormonal changes mean that some fetuses with a Y chromosome can still develop as female, while others are born with abnormalities in their genitals that make it unclear whether they are biologically male or female (known as intersex).

When it comes to gender – how people identify themselves in terms of masculine and feminine – or sexuality (those they are attracted to), the situation becomes even more complex and controversial. It is now increasingly accepted that the concept of gender is drawn from many elements, including a person's psychology, physical appearance and underlying genetic makeup, as well as social cues and expectations. Based on the results of studies comparing identical and non-identical twins, scientists also think that genes help to shape someone's sexuality,

The Case of the Calico Cats

Calico cats (also known as tortoiseshells) are thought of as good-luck charms in many cultures all over the world. Their blotchy coat colouring is also a good example of X-inactivation at work. There are several genes responsible for coat colouration in cats and these are carried on different chromosomes. One of them – found only on the X chromosome – exists in two different versions (alleles), creating either black or orange fur. Virtually all calico cats are female, so they randomly switch off one or other of their two X chromosomes as the kitten grows in the womb. If a female cat inherits both orange and black fur alleles – one on each X chromosome – then these will be randomly switched off in all the hair-growing cells, creating the characteristic colour blotches.

Because male cats have just one X chromosome, they only have one of the coat-colour alleles – orange or black. Because their X chromosome stays active, this gene will be switched on in all their fur, resulting in a ginger or black tom. Male calico cats do exist but are extremely rare. They usually turn out to have two X chromosomes, one with the orange fur allele and the other with the black, in addition to a Y chromosome. Even though the Y chromosome makes them male, they will still undergo random X-inactivation in all their cells and end up with a calico coat.

although more influences are brought from the environment and upbringing. In this chapter we will refer to genetic sex determination as "sex", with the assumption that XX is female and XY is male, acknowledging that this is not the case for everyone and that a person's gender does not automatically follow their biological sex.

MAKING A MALE

Until around the sixth week of pregnancy, there is no detectable difference between human male (XY) or female (XX) embryos growing in the womb. They both have two sets of tube-like structures – the Wolffian and Müllerian ducts, and little pockets of germ cells (the cells that will go on to make eggs or sperm) – running through the lower body. The critical moment in making a male occurs when a gene on the Y chromosome, called "SRY", is switched on.

SRY encodes a transcription factor that switches on a cascade of genes that make some important changes. The Müllerian ducts break down, and the Wolffian ducts start growing into the internal plumbing of the male reproductive system. Then the pockets of germ cells start to grow into testes, producing male hormones and activating genes that create further male changes within the developing fetus. In the absence of SRY – in either XX female embryos or XY males with a faulty version of the gene – the opposite occurs as the default pathway. The Wolffian ducts break down, the Müllerian ducts grow into the female reproductive system, female-specific genes are activated, and the germ-cell pockets become ovaries.

The XX and XY chromosome system of genetic sex determination is the same in mammals and other animals, including fruit flies. There are other ways of doing it, though. Birds have W and Z sex chromosomes, with the male being ZZ and the female ZW. Some organisms, such as worms, are hermaphrodite and so have both male and female sex organs. In reptiles, the sex of some species depends on the temperature at which their eggs are incubated during a crucial time in development. Some animals, such as fish and snails, can even change sex during their lifetime.

SWITCHING OFF

The fact that female mammals – including humans – have two X chromosomes, while males have just the one, means that one sex has a double dose of X chromosomes compared with the other. In 1961, British geneticist Mary Lyon published a paper in the journal *Nature* showing that female mice solve this problem of double dosage by randomly switching off one or other of their two X chromosomes in all cells of the body, silencing all the genes on that chromosome. We now know that humans and all other mammals do the same thing. This process – originally called Lyonization after Lyon's work, but now known as X-inactivation – begins when the embryo consists of just a handful of cells.

The main player in X-inactivation is a gene called "XIST", and this is just found on the X chromosome. Importantly, XIST is only switched on in cells with more than one X chromosome – usually in all the cells in an XX female embryo, and none of the cells in an XY male. XIST is only transcribed from one of the two X chromosomes – the one that will be switched off (inactivated). Unlike most of the genes on the X chromosome and elsewhere in the genome, XIST does not make a protein. Instead, it produces a very long non-coding RNA that is around 17,000 "letters" long. Many copies of XIST RNA wrap themselves around the X chromosome that is due to be inactivated, forming a kind of RNA coat. In turn, this recruits silencing proteins and epigenetic marks (*see* Chapter 9) to keep the genes switched off. The decision to inactivate one or other X chromosome is a random choice, but it is inherited as cells divide and multiply. This creates a patchwork made up of clusters of cells that have all inherited the same inactive X, technically called a mosaic.

DOUBLE TROUBLE

We inherit two copies of each chromosome – one each from both our parents. For the 22 pairs of regular autosomes, this means that there are two copies of every gene in each chromosome pair. If there is a mistake or fault within a gene on one chromosome, then the copy on the other chromosome can act as a back-up and keep the cells working properly. However, this process does not work for the X and Y chromosomes. Males only have one X chromosome, with a different set of genes on the Y, so inheriting a fault in an X chromosome gene means there is no back-up. This is the cause of a number of diseases and traits known as X-linked conditions. Perhaps the most famous of these is haemophilia, which affected the male descendants of Queen Victoria for three generations (see Chapter 5).

Another good example of an X-linked trait is red/green colour blindness, which affects roughly one in 12 men and one in 200 women. It is the result of inheriting a mistake in either of two genes on the X chromosome – "OPN1LW" or "OPN1MW" – which encode light-sensitive pigments within the eye that enable us to distinguish different colours. A female inheriting just one faulty version will have normal vision – even though one of the X chromosomes is inactivated in female cells, the remaining functional copy is enough to compensate. On the other hand, if an XY male has a faulty copy of one of these genes on his sole X chromosome, he will definitely be colour blind, as there is no genetic back-up. Colour blindness is much less common in women, as they have to inherit two X chromosomes carrying a faulty gene to have the condition. However, unaffected women carrying one copy of an X-linked disease gene can still pass it on to their children, and their sons will have the condition if they inherit that particular X chromosome.

Although X-linked conditions are much less common in females than males, unusual conditions can still arise from the mosaicism caused by X-inactivation. In 1901, the German dermatologist Alfred Blaschko noticed that certain skin conditions all seemed to follow the same patterns: sweeping lines running across the chest and back and down the arms and legs, similar to the stripes on a tabby cat or brindled dog. These lines correspond to the paths taken by skin cells as a fetus grows in the womb, multiplying and spreading out along the torso and limbs. In most people, Blaschko's lines are completely invisible, but in females with faults in certain genes on one of their two X chromosomes, the random patterns of X-inactivation mean that some groups of embryonic cells end up activating the gene, while others do not. This can cause stripes of genetically different skin cells across the body. If the fault is in a skin pigment gene, then these stripes will be two different colours, leading to unusual stripy skin patterns.

Genetic mosaics can also occur as a result of alterations to genes on regular (autosomal) chromosomes, particularly if a change happens very early in embryonic development. In this case, a random change (mutation) affects a gene in just one of the cells in an early embryo. Sometimes this change could be dominant, meaning that a modification in just one copy of the gene is enough to cause an effect. Alternatively, it could be a mutation that affects the remaining functional copy of a gene

---

Mum Versus Dad

We have two copies of all our genes, except those on the X and Y chromosome. For most of them, if both copies are functional then both copies are usually active. This is not the case for some genes, known as imprinted genes. Depending on the gene, either the mother *or* the father's copy is active and the other one is silenced. These imprinted genes carry special patterns of DNA methylation (*see* Chapter 9), which are put on when eggs and sperm are made. For example, only the paternal (father's) copy of a gene called "Igf2" is active during development, while the copy inherited from the mother is marked with DNA methylation and is always switched off. Igf2 makes a protein that promotes the growth of cells – it must be carefully controlled so that a fetus doesn't grow too large in the womb.

There are probably a couple of hundred imprinted genes in the human genome and many play important roles in growth and development, particularly in the brain. Mistakes in imprinting can lead to conditions such as Prader-Willi syndrome (PWS) and Angelman syndrome (AS), which are due to loss of, respectively, either the paternal or maternal copies of certain genes on human chromosome 15 and have almost opposite appearances. Children suffering from PWS are constantly hungry and become obese while AS children have difficulty feeding, among other physical and mental-health problems.

---

in an embryo that is already carrying a faulty copy of that gene in all of its cells.

Either way, as the embryo develops, the faulty cells will grow and divide, taking their place in the growing fetus. Sometimes the effects of this mosaicism, such as patches of oddly pigmented skin or hair, are not serious, while at other times they can be much more so. Mosaic changes in a gene called "AKT1" cause patches of cells to grow out of control, leading to unusually large body parts – a condition known as Proteus syndrome, after the shape-changing Greek god.

ADAM AND EVE

Although the biblical story of Adam and Eve is a myth, genetic research tells a fascinating tale about our early ancestors. The Y chromosome is always handed down from father to son, but it does not change much from generation to generation. This makes it relatively easy to trace male genetic ancestry and draw up a kind of family tree for groups of people with similar Y chromosomes, known as haplotypes. By studying the differences between these haplotypes and factoring in how genes have changed through time, geneticists think that the last ancestor to share a Y chromosome with all men alive today – the so-called "Y-chromosome Adam" – lived around 200,000 years ago.

Some researchers use Y-chromosome haplotype information to map the spread of different populations around the world, particularly within Europe. This can be refined to a surprising level of detail. In the UK, it is customary for a father's surname to be passed on to his children, so any sons will inherit his Y chromosome as well as his name. At the University of Leicester, Professor Mark Jobling and Dr Turi King looked at whether British families who share the same surname also have the same Y chromosome haplotype. They studied more than 1,600 men with 40 different British surnames and found that those with relatively unusual names – such as Grewcock, Wadsworth, Ketley and Ravenscroft – tended to have similar Y chromosomes. Tracing this back in time suggests that each family may have come from a single ancestor with that name around 700 years ago. The research also showed that people with common surnames, like Smith or Baker – which were usually used in the past to describe people's occupations – were not more likely to share a similar Y chromosome than randomly selected men within the UK population.

Unlike the Y chromosome, which passes almost unchanged from father to son, X chromosomes become much more mixed up as they pass down the generations, especially during meiosis when egg cells are made (see Chapter 4). But there is another way of tracking back through the female line: mitochondrial DNA. All human cells contain molecular power stations called mitochondria, which generate the energy that cells need for life. Unlike other parts of the cell that rely on instructions from genes in the cell's nucleus, mitochondria have a very small amount of their own DNA – just 37 genes. Egg cells are full of mitochondria while sperm contribute none to the embryo, so all the mitochondria in a baby's cells will have come from its mother. Like Y chromosome DNA, mitochondrial DNA changes relatively slowly over time and is only passed down the maternal line. By comparing mitochondrial DNA from people from across the world, researchers believe that the last common female ancestor with whom we all share our mitochondrial DNA (the "mitochondrial Eve") lived around 150,000 to 200,000 years ago.

The concept of this genetic Adam and Eve is a bit misleading, as it is impossible to put a precise date on the time when they were alive. It certainly does not mean that our species arose from just one man and one woman, either. Rather, these individuals happened to be the one man and woman – out of many alive at the time – whose Y chromosome and mitochondrial DNA made it all the way down the generations to today's humans.

So far we have focused a lot on the genes that build our brains, bodies and blood. But recent research has shown that long-dead viruses within the human genome have also played a vital role in shaping our species.

CHAPTER FIFTEEN

THE VIRUSES

THAT MADE US

HUMAN

**Viruses are not just germs that make us
sick. They have also played a vital role in
the evolution of the human genome.**

Viruses are best described as genetic information – usually RNA – packaged in a protein coat. Unlike bacteria, which can live and multiply on their own, viruses can only replicate inside a host cell. The latest estimates suggest that there are more than 300,000 different types of virus that can infect mammals, and many more that infect other species. Some of them, such as influenza (flu) or Ebola viruses, enter our cells and hijack the cellular machinery to turn out millions of new viruses that go on to infect more cells and other people. One class of virus is more sinister, though.

Known as retroviruses, these viruses convert their RNA-based genetic information into DNA by using an enzyme called reverse transcriptase. This viral DNA then inserts itself into the genome, lying dormant until the time is right to start making new viruses again. HIV (human immunodeficiency virus) is probably the best example of a retrovirus that infects humans, but there are others out there too. Because retroviruses can smuggle themselves into our DNA, and if this occurs in a germ cell (egg or sperm), the viral DNA can be passed on to the next generation. Over time, these viral DNA sequences pick up mutations that render them unable to function properly or make new viruses to infect other cells. They're effectively dead, but they remain in the genome.

Since the development of large-scale DNA sequencing technology, scientists have discovered many viruses that found their way into the human genome thousands (or even millions) of years ago. Some estimates suggest that up to 80 per cent of our DNA might have originally come from viruses in some form or other. Rather than just being long-dead genetic fossils, some of these viral sequences have become very useful to us. In some cases, they have played a vital role in shaping the evolution of our species.

MEET YOUR HERVS

"Dead" retroviruses that embedded themselves in the genome but lost the ability to make new viruses are known as retroelements. Up to 40 per cent of our genome is composed of repeated retroelements. Some of these sequences are probably so-called junk DNA (*see* Chapter 1), but others are useful. One important group contains the human endogenous retroviruses, or HERVs, which make up about 8 per cent of the human genome and occupy several times as much DNA as our actual protein-coding genes. There are many different types of HERV scattered through our genome, and they all originally came from viral infections long ago in our evolutionary history. Many are coated with epigenetic marks, such as DNA methylation, that keep them quiet and out of trouble (*see* Chapter 9), but sometimes they can evade this silencing and become active.

Although they are no longer infectious, HERVs can still be transcribed by RNA polymerase to make RNA, which may then be translated into proteins or even converted into DNA. HERVS also contain sequences that act as genetic switches to activate nearby genes. In 2016, scientists at the University of Utah made a fascinating discovery. They found that HERV sequences called MER41, which originally came from a virus that entered our genome between 45 and 60 million years ago, are responsible for switching on genes that respond to a molecule called interferon. This is the primary danger signal – it tells immune cells to gear up and fight off viral infections. There is also evidence that DNA made from HERV RNA is involved in provoking an immune response against viruses. These ancient viral sequences have actually become genetic double agents, helping cells to tackle other viruses that are trying to invade.

Researchers in Singapore have found that one type of HERV, called HERVH, is actively transcribed into a long non-coding RNA in the stem cells buried within a very early human embryo. Furthermore, the HERVs themselves seem to act as control switches, turning on genes involved in maintaining the special properties of these embryonic stem cells. HERVH is not active within any other types of cells in the body, and is only found in humans and our closest ape cousins – gorillas, apes, orangutans and bonobos. It is a good example of how primates have effectively domesticated this virus to help out during early development.

Even more strangely, researchers led by Professor Joanna Wysocka at California's Stanford University have discovered that one particular HERV is reactivated when the embryo is just a tiny ball of cells, making virus-like particles that stay inside the embryonic cells. It is not entirely clear what these trapped viruses are doing, although they may play a role in helping to protect embryos against external harmful viruses that might be trying to infect them. Again, these ancient viruses in our genome have been harnessed to help shape the way our cells work today.

As well as our HERVs, there are other types of retroelement in the human genome. Some of these, known as retrotransposons, make up much of our repetitive "junk" DNA, and were thought to be long-dead genetic fossils. New research, using highly sensitive DNA sequencing techniques, has revealed that these retrotransposons can start moving

Barbara McClintock and the Jumping Genes

American cell biologist Barbara McClintock was the first person to discover so-called jumping genes (transposons), through detailed experiments with corn plants in the 1940s and 1950s. She noticed that certain genetic elements could hop around in the genome of the plant, causing unusual colour patterns on the corn kernels. When McClintock first made her groundbreaking discovery, very few researchers believed the results because they thought the genome was fixed and static. She struggled to publish her findings or receive recognition for her ideas, but the importance of her work to our understanding of how genes function, including the role of jumping genes, was later recognized. As a fitting tribute, in 1983 she was finally awarded the Nobel Prize in Physiology or Medicine at the age of 81.

about within the genome, particularly within nerve cells in the brain. Again, scientists are not clear about what these jumping sequences are doing, but they may be important for increasing both the genetic diversity of neurons and their computing power. On the more negative side, they could also play a part in psychiatric conditions, such as schizophrenia, or brain changes associated with ageing. In addition, HERVs and other retroelements are implicated in cancer and autoimmune diseases (whereby the immune system attacks healthy cells in the body) by disrupting or affecting the activity of crucial genes.

THE VIRUSES THAT MADE MAMMALS

Around 15 years ago, researchers in Massachusetts discovered a new human gene that was only switched on in cells in the placenta – the organ that connects a growing fetus to the womb in mammals. Called "syncitin", it encodes a protein that causes cells to join together, creating a layer of large fused cells that make up an important part of the placenta. Curiously, the DNA sequence of syncitin looks very much like the code used by some retroviruses to make their protein coat. Later, other scientists found another human syncitin gene, which is also involved in making the placenta and helps to dampen down the mother's immune system, preventing it from attacking the growing baby. Again, the gene looks very much like a retrovirus. Clearly, humans adopted these viral genes to do a vital job in development millions of years ago, and they became a permanent fixture of our genome.

Intriguingly, while humans and our primate cousins share these two syncitin genes, they are not present in any other mammals, despite the fact that these animals also have placentas with a layer of fused cells. It is now known that mice also have two syncitin genes. Although these do the same job as ours, they resemble completely different viruses. In addition, there is yet another distinct, virally derived gene playing an identical role in cats and dogs, which are descended from the same ancestral carnivores. However, pigs and horses do not have the layer of fused cells in their placenta, nor do they have any genes that look like other mammalian syncitins. Perhaps they never caught the right virus far back in their evolutionary history.

CHANGING FACES, CHANGING BRAINS

As well as looking at the role of HERVs in early embryos, Stanford University's Professor Joanna Wysocka is searching for the key genetic differences between humans and chimpanzees. Although our genes are almost identical to those of chimps, we certainly do not look similar or behave in the same way. The crucial differences must lie in the control switches that turn genes on or off during development. Wysocka and her team are focusing on a cluster of cells called the neural crest. These turn into a range of cell types as a fetus grows in the womb, including pigment cells in the skin (melanocytes), supporting cells in the brain, some types of muscle and nerve cells, and also the bones and cartilage of the face.

Both chimps and humans have neural-crest cells, and they do the same job in the developing fetuses of both species. However, the shapes of our faces differ a great deal to those of our chimp cousins. Wysocka decided that this must be due to distinct patterns of gene activity within human neural-crest cells compared with those of chimps. When she looked closely at the activity patterns of genes and the switches controlling them, she found that many of the switches that were active only in human neural-crest cells looked like they were made from retroelements. These ancient viral remnants probably played an important role in creating our flat, delicate human faces out of the chimp-like, long-snouted countenances of our ancestors.

Another example of a virus that has helped to shape our species is found near a gene called "PRODH", which encodes a protein that sends signals between nerve cells in the brain. It is particularly active in a region called the hippocampus that is involved in making memories, thanks to a control switch made from a long-dead endogenous retrovirus. Chimpanzees also have a version of PRODH, but as they do not have this particular virus sequence nearby, they do not make nearly so much PRODH protein in their brains.

One possible story is that a mutation in our ancient human ancestors somehow resulted in a copy of the viral DNA next to PRODH and it started activating the gene in the brain. This genetic change did not occur in chimps, so they do not have the same level of gene activity. It is unclear how this alteration has influenced our brains, compared with

Bacterial Buddies

More and more scientific and media attention is now focused on bacteria, another type of microbe living inside us. Early estimates suggested that we may have as many as ten times more bacterial cells in our bodies than human cells, but the latest calculations suggest that this ratio is closer to one to one. Even so, that adds up to trillions of bugs living in our guts, on our skin and elsewhere. Rather than just being freeloaders, our personal bacteria – known as the microbiome – are now credited with playing an important role in our health and wellbeing. Scientists are currently unpicking the relationship between gut bacteria and all kinds of conditions ranging from obesity to bowel cancer, and even mental-health problems. One day, it might be possible to manipulate the different species of bacteria living within us to boost health, control weight or treat disease, but a lot more work still needs to be done before this can become a reality.

chimps, but as faults within PRODH could be involved in brain disorders such as schizophrenia, it is likely to have some kind of significance.

New research comparing the DNA sequences of many different species of animals is now revealing that long-dead viruses make up many of the control switches in the genome, turning genes on at the right time and in the right place. These switches tend to differ somewhat between species, while genes are more similar. For example, one study from Canadian researchers showed that most of the control switches that are only found in primates (and not in other mammals) look like they originally came from viruses. Perhaps specific viral infections in groups of ancient mammalian ancestors helped to shape their evolution into different species.

It is clear that the viruses we picked up many millennia ago have played a vital role in the evolution of the human genome, and still control many aspects of our genes and cells today. Maybe the viruses infecting our species now will become important aspects of our genome in the future. Of course, it is impossible to tell as it takes millions of years to trap and harness these wild viruses within the confines of our DNA.

It is not just viruses and bacteria that can make us ill by causing infections. Changes in our DNA also have a big impact on health, as we will see in the next chapter.

CHAPTER SIXTEEN

WHEN THINGS GO WRONG

Nothing lasts forever, including the human body, and our genes play an important role in the diseases that affect us as we age.

It is a sad fact of life that every one of us will eventually die, although we all hope for as long and healthy a life as possible. Advances in public health – such as better sanitation, good nutrition and vaccines – have meant that life expectancy has risen dramatically in many countries over the past century. However, with a longer lifespan has come a greater incidence of diseases that tend to strike in old age, such as cancer and dementia.

Advances in early diagnosis and treatment mean that around half of all those diagnosed with cancer today will live for at least ten years, and average survival has doubled since the 1970s. There is still a long way to go before we can really cure cancer, and the reason it is such a hard-fought battle lies in our genes.

OUT OF CONTROL

Cancer starts when a single cell – or maybe a small group of cells – begins to multiply out of control, forming a tumour. Eventually, the cancer cells move away from this primary tumour and spread through the bloodstream. They set up new secondary tumours elsewhere in the body, especially the brain, bones, lungs or liver. Because genes are the instructions that tell our cells to multiply when they are needed and die when they are damaged, any changes to key control genes will increase the chances of developing cancer and enable it to start spreading.

After more than a century of dedicated research, we now know a huge amount about the genetic changes that underpin cancer. This has accelerated rapidly in the past few years, thanks to the development of large-scale DNA sequencing technology. This enables scientists to read the genomes of thousands of tumour samples from patients all over the world. Although some hereditary gene faults boost the risk of cancer – such as changes within "BRCA1" and "BRCA2" that are linked to breast, ovarian and prostate cancers – as well as more subtle inherited gene variations that have a smaller impact, most genetic changes that drive cancer accumulate over a lifetime. This is why the risk of cancer increases as we age – there has simply been more time for us to pick up mistakes in our DNA (see Chapter 4 for more on the causes of DNA damage). However, there is no single, crucial "cancer gene". Instead, scientists think that a series of faults in key genes needs to build up for cancer to really get going.

There are two main groups of genes that are relevant to cancer. The first was discovered during the 1970s when scientists studied the viruses that made cells become cancerous. They realized that the genes in the viruses that were making the cells grow were actually copies of normal genes involved in cell division. Somewhere along the line during evolution, the viruses picked up these genes, causing cells to grow out of control when they infected them. Scientists later discovered that many human cancers had faults in their own normal versions of these cell division genes – which are known as oncogenes – rather than needing a virus to kick-start the process. Until this point, it was believed that viruses cause all cancers, but this new discovery showed that cancer could be started by genetic faults within our own cells. Mistakes that activate oncogenes are acknowledged as major drivers of cancer, causing cells to multiply when they should not.

Many oncogenes encode proteins that send or receive signals coming into cells telling them to divide. Normally, these signals are only sent when new cells are needed – for example, to replace dead or damaged cells or as an organism grows. Certain mutations in oncogenes mean the signalling proteins they make are permanently switched on, continually telling a cell to divide, even when there are no signals coming in. They can be considered similar to an accelerator in a car that is pressed down, making it go faster and faster.

The second significant group of genes are known as tumour suppressors and they act as the cell's brakes. Unlike oncogenes, which

A Lethal Weapon

Most chemotherapy drugs work by stopping tumour cells from multiplying. However, they are not specific to cancer and prevent healthy cells from growing too, causing side effects such as sickness, blood problems and hair loss. The hunt is on to find kinder, targeted treatments that specifically attack cancer cells. One exciting approach is synthetic lethality, pioneered by UK scientists Professors Steve Jackson and Alan Ashworth. They realized that cells have two ways of repairing damage to their DNA: one involves proteins made by the BRCA1 or BRCA2 genes (see Chapter 4); another uses a protein called PARP (short for poly-ADP-ribose polymerase).

Cancer cells lacking BRCA1 or BRCA2 (growing in people who have inherited a faulty version of one of these genes) have to rely on PARP to repair damage to their DNA. Professors Jackson and Ashworth, and their colleagues, found that PARP-blocking drugs (known as PARP inhibitors) would leave these cells with no way to repair their DNA, so they would die. The idea was successful. In 2015, the drug olaparib (Lynparza) became the first PARP inhibitor to be approved for treating ovarian cancer in women with a faulty BRCA gene. The search is now on for other combinations of DNA repair genes that could be targeted in the same way.

cause cells to multiply, tumour suppressors protect us from cancer. They make proteins that are involved in processes such as DNA repair or cell death, and are responsible for fixing or killing off damaged cells before they can develop into a tumour. Inactivating faults in tumour-suppressor genes mean that cells do not repair themselves properly or die when they should. For a cell to start multiplying out of control and become cancerous, it needs to have a stuck accelerator (overactive oncogenes) and no brakes (inactive tumour suppressors).

TRUNKS AND BRANCHES

Large-scale genome-sequencing studies of tumours are revealing a huge amount of information about the genetic changes that underpin each individual patient's disease, and we now know that every person's cancer is unique. In many cases there may be hundreds, or even thousands, of mutations in each tumour, although not all of them will be responsible for making the cancer cells grow and spread (so-called driver mutations). Some changes, known as passenger mutations, are irrelevant and just along for the ride. This poses a big challenge for doctors, because we need to move away from the idea that cancer should be defined by the place where it starts in the body, and think instead about targeting specific gene faults that drive each person's tumours.

Many pharmaceutical companies are developing drugs that block specific overactive signalling molecules produced by oncogenes, although they will only work for people whose cancer is driven by that particular gene fault. These are known as targeted therapies. Working out exactly which targeted drugs will work for which patients – depending on their genetic makeup and the gene faults driving their cancer – is known as precision or personalized medicine. This is a rapidly growing area, and while many hopes are pinned on this approach leading to new cures, it really is not that simple.

Just as Darwin showed that organisms adapt in response to changes within their environment, leading to evolutionary changes over time, cancers evolve within a patient's body on a much shorter timescale.

Because cancer cells have faulty DNA-repair systems, they can quickly adapt to withstand DNA-damaging chemotherapy and radiotherapy. They can also evolve more changes in signalling genes, and so no longer respond to targeted drugs. Any little pockets of tumour cells that are resistant to a treatment will survive and keep growing and changing, so the disease can return months or even years later – only this time, the cancer cells will all be resistant and the treatment will fail.

Tumour evolution is a big problem, and is the main reason cancer is so difficult to cure. Researchers are trying to pinpoint any rules or patterns that explain how cancers evolve and develop resistance to treatment. If they succeed, this would be a major step forward in understanding cancer and tackling it effectively.

BRAIN BREAKDOWN

Cancer is not the only disease we need to worry about as we live longer. The risk of dementia increases with old age, roughly doubling every five years from age 65 to 90. There are many types of dementia, but they are all characterized by progressive loss of memory and brain function, along with changes in behaviour and personality. The most common form is Alzheimer's disease, named after Dr Alois Alzheimer, the German psychiatrist who first described it in 1906.

We now know that Alzheimer's disease has dramatic effects on the brain. One obvious sign is the accumulation of harmful forms of two proteins called amyloid and tau. These build up in the brains of people with the condition, creating damaging, clump-like amyloid plaques and long tangles of tau. At the moment, scientists are not sure whether amyloid or tau is the main driver of Alzheimer's disease, or indeed if there is another underlying process at work.

Drugs aimed at eliminating either protein have not been successful in clinical trials. This may be because, by the time that plaques and tangles have started to grow, it is too late for treatments or preventative drugs to work. So there is also a need to find ways of diagnosing the disease much earlier. In 2016, the UK Medical Research Council launched a major study among 250 people who may be at risk of developing Alzheimer's disease, monitoring them with a wide range of tests and brain scans in the hope of spotting the very earliest signs. The team, led by researchers at the University of Oxford, hopes to spot physical changes in people who are starting to develop the condition, so that treatments can be tested much earlier.

Why Don't Elephants Get Cancer?

During the 1970s, British biologist Professor Sir Richard Peto noticed an unusual paradox – if the chance of a single cell in the body becoming cancerous is the same across all animals, then larger, longer-lived species, such as elephants, should develop many more cancers than humans, as their bodies contain many more cells. Yet elephants hardly ever get cancer, and in 2015 Dr Josh Schiffman and his team at the University of Utah discovered why. By studying DNA samples from elephants in Utah's Hogle Zoo, they found that the animals have extra copies of the "p53" guardian gene in their genomes. The extra dose of p53 gives them even stronger protection against cancer, ensuring that any damaged cells will die rather than continue to multiply and form tumours.

Another animal that is unusually resistant to cancer is the naked mole rat – a strange-looking, hairless rodent species that lives underground in the African desert. These creatures have special tumour-suppressor genes that help to protect them against cancer. They also possess an unusual version of a gene called "hyaluronan synthase 2", which makes a very large, sticky molecule. This acts as a kind of cellular glue, preventing any cancer cells from spreading throughout the mole rat's body.

New clues for diagnosis and treatment may also be uncovered from studies searching for gene variations that increase the risk of developing Alzheimer's disease. By examining families affected by a hereditary form of Alzheimer's, starting in mid-life, researchers have found faults in three key genes that seem to be important. One of them is called "APP" and this encodes a protein that becomes amyloid. The others are proteins known as presenilin 1 and presenilin 2 (made from the "PSEN1" and "PSEN2" genes), which are involved in cutting up the APP protein, normally preventing the build-up of harmful, plaque-forming amyloid.

In the Blood

When it comes to treating cancer, one of the biggest challenges lies in monitoring how well it responds to treatment and whether the tumour cells are becoming resistant to the therapy. At the moment, this is done using CT and MRI scans, or by taking tumour samples (biopsies) with surgery. Scientists are now developing methods of tracking DNA or whole cancer cells that are shed into a patient's bloodstream as their tumour breaks down, analyzing genetic changes using sensitive DNA-sequencing techniques. Known as liquid biopsy, this new technology enables doctors to see within a matter of days whether a treatment is working, and also follow how the cancer is evolving and changing at a genetic level. Perhaps one day we will even be able to diagnose cancer via a simple blood test, using information from tumour DNA in the bloodstream to predict the most effective treatments for each individual patient.

If someone inherits a faulty version of one of these three genes, they have a high risk of developing Alzheimer's disease during their thirties or forties. Further studies among the general population have found a handful of genes with smaller effects on the risk of developing the condition in older age. Some of them are involved in inflammation – a type of immune response – while others play a role in clearing amyloid out of the brain. The strongest association with risk comes from a gene called "APOE", which makes a molecule called apolipoprotein E. There are three versions (alleles) of APOE in humans – confusingly numbered 2, 3 and 4. Inheriting two APOE4 alleles (one from each parent) increases the risk of developing Alzheimer's disease by 12 times more than the least risky version, while carrying just one copy of APOE4 increases the risk fourfold. Because so much attention has been focused on amyloid and tau, nobody is quite sure what apolipoprotein E does within the brain, or how variations in the gene increase the risk of Alzheimer's disease.

This is just one form of dementia, although there are also many lesser-known varieties. Other conditions that affect the brain and nervous system, such as Parkinson's and motor-neurone disease, are poorly understood and currently incurable. Unfortunately, research into dementia and other neurodegenerative diseases is still relatively underfunded, despite their being a growing problem within the ageing population.

We have taken a detailed look at how our genes work, how they build our bodies, and what happens when they go wrong. Finally, it is time to ask what will happen to our genes – and our species – in the future.

CHAPTER SEVENTEEN

HUMAN 2.0

**Designer babies, replacement organs or
even extinction – what is the next step in
our evolutionary journey?**

The human species has been shaped by millions of years of evolution, from our very earliest mammalian ancestors to our most recent Neanderthal cousins. We are still evolving today – as are all the other species on the planet – although these changes are taking place too slowly for us to see. As we saw in Chapter 2, studying DNA from living people all around the world, as well as long-dead fossils, can build up a picture of our evolutionary journey so far. What it cannot tell us is where we will end up in the future.

Since the discovery of the structure of DNA in the 1950s, scientists across the world have developed ever more advanced ways of sequencing and studying genes, along with powerful tools for genetic engineering. This means that we are the first species to have developed methods of deliberately and precisely altering our genes. While these techniques could help to cure diseases, reduce suffering and prolong life, they also raise big ethical questions. What alterations are acceptable? Who should benefit from them? Should we be permanently changing the human genome, potentially altering the genetic trajectory of our species into the future?

OLD GENES, NEW TOOLS

Genetic engineering is nothing new. Humans have been selectively breeding animals and plants for thousands of years. All the domestic and agricultural animals and plants we have today have been genetically engineered by choosing and breeding individuals with desirable traits, such as gentleness in dogs, more muscular beef cattle, or high-yielding wheat crops. In some cases, crop breeders have even exposed seeds to radiation – so-called atomic gardening – in the hope of generating plants with useful mutations that can then be bred into new strains. None of these approaches are very precise, and breeding organisms for a beneficial trait can bring a less useful characteristic along for the ride. For example, many Dalmatian dogs have hearing problems, as the gene variation that gives them their elegant spotty fur also affects important pigment-making cells inside their ears.

It was not until the 1960s and early 1970s that scientists developed tools enabling them to tweak genes in a targeted way by cutting and pasting different pieces of DNA together. Within a few years, this had led to the production of genetically modified bacteria that produce human insulin, which is now used as a life-saving treatment for diabetic people all over the world. By 1981, scientists had created the first genetically modified mice that could pass entirely new characteristics to their offspring. Although these techniques were much more accurate than selective breeding or generating random mutations, allowing researchers to specifically add in or knock out genes to see how they worked, they were fiddly, expensive and time-consuming.

Then in 2012 came a major breakthrough. Professor Emmanuelle Charpentier at Umeå University, Sweden, and Professor Jennifer Doudna, her collaborator at the University of California, Berkeley, looked at how bacteria respond to virus infection by making molecular scissors (called Cas9) that cut up invading viral DNA. They were able to target these scissors to any sequence of DNA by using specially designed short fragments of RNA known as CRISPR (clustered regularly interspaced short palindromic repeats), making precise changes quickly and easily. Although Doudna and Charpentier's work was done using DNA in a test tube, other laboratories – including that of Professor Feng Zhang and his team at MIT in Boston – were quick to figure out how to apply the CRISPR/Cas9 system to living cells.

Since then, researchers all over the world have started to use CRISPR (as this is usually known) to make highly targeted alterations to genes in all kinds of organisms, including humans. In 2016, scientists at the Salk Institute in California announced that CRISPR could repair genes in adult rats suffering from a genetic form of blindness. Another team fixed a faulty gene causing liver damage in mice, paving the way for similar approaches in people. In the same year, researchers in China revealed that they had used the technique to modify a cancer patient's immune cells to make them more effective at destroying tumour cells.

Several companies are now working on CRISPR-based treatments for a range of diseases. One approach is to fix faulty genes within blood stem cells in people suffering from disorders such as anaemia and thalassaemia, which are caused by faulty genes. Another is to modify liver cells to act as miniature factories, pumping out proteins that are missing in people with certain genetic conditions. Perhaps the biggest challenge lies in working out how to deliver CRISPR into cells within the body, although there are some promising results using nanoparticles and viruses. Even so, it is likely that there will be many more breakthroughs and treatments based on this technique in the future.

DESIGNER BABIES

In July 1978 a very special baby was born in Oldham, a town in the north of England. Louise Brown was the first child conceived through *in vitro* fertilization (IVF), making her the first of more than a million test-tube babies that have since been born all over the world. Further developments and improvements in IVF technology – such as preimplantation genetic screening and diagnosis (*see* Chapter 5) – have brought joy to many families, but also sparked heated ethical and legal debates, especially around the area of genetic modification.

In 2015, scientists in China used CRISPR to modify genes in discarded, non-viable human embryos generated through IVF, proving that it could be done. Given that the tools required to genetically modify human embryos exist (and are currently being tested), there is an ongoing discussion about whether or not they should be used to fix human embryos carrying inherited genetic faults that lead to severe diseases (so-called Mendelian diseases, *see* Chapter 6).

For families affected by these diseases, genetic modification could eradicate the faulty genes that have plagued them. Other people see it as playing God – involving unacceptable risks and permanently altering the human genome in a way that would be passed on down the generations. The decision about whether to press ahead with research in this area needs to be made by society as a whole – including scientists, doctors, patients, ethicists, lawyers and the general public – and it may not be possible to get every country in the world to agree to the same rules.

One significant step further forward is the concept of designer babies: using genetic engineering to create embryos that carry certain gene variations for characteristics such as eye colour, intelligence, height and so on. This is going to prove much trickier, if indeed it is possible at all. Most traits cannot be pinned on a few genes or control switches, so working out which ones to tweak will be very difficult. As we have seen in other chapters of this book, the relationship between someone's genes and how they turn out is not at all straightforward. Genes play a role, of course, but the environment and upbringing also have an impact, as do the effects of epigenetic modifications. We are therefore unlikely to see true designer babies at any time in the near future.

However, genetically modified humans already technically exist. In 2001, US scientists announced the birth of the first babies that carry DNA from three parents, through a technique called cytoplasmic donation. These children were created for families suffering from mitochondrial disease – a condition where the power stations inside cells, called mitochondria, do not work properly. As we saw in Chapter 14, every baby's mitochondria comes from the mother's egg cells and contains a very small amount of their own DNA. Babies born to women with faults within this handful of mitochondrial genes suffer many health problems, and usually die very young.

Cytoplasmic donation involves taking some cytoplasm from a healthy donor egg (containing mitochondria with normal DNA) and injecting it into the egg of the mother with faulty mitochondria, which is then fertilized in vitro with the father's sperm. This technique is no longer used, but a new approach – known as mitochondrial donation – is now being tried. In 2016, US doctors working in Mexico announced the birth of the first baby to be conceived using this method. Mitochondrial donation is illegal in many countries, although it was recently approved in the UK. Some people are concerned that it may cause unforeseen problems, especially as the donor mitochondrial DNA will be passed on to any descendants if the baby is a girl. However, families affected by this devastating disorder are hopeful that it will enable them to have healthy babies in the future.

STEM CELLS AND SPARE PARTS

Another controversial aspect of our future genetic possibility is human cloning. In 1996, researchers at the Roslin Institute in Edinburgh, led by Professors Keith Campbell and Sir Ian Wilmut, successfully created Dolly the sheep by placing the DNA from an adult sheep's breast cell into an egg cell from which the DNA had been removed. Dolly was the first mammal to be cloned from an adult cell – something that was previously thought to be impossible. Many different species of mammals have been cloned since then and scientists in several countries, including South Korea and the UK, have made early human embryo clones. At the moment, these embryos can only be grown for 14 days in the laboratory and cannot be implanted into the womb to develop further.

Scientists were initially very excited about human cloning, not for the possibility of creating cloned babies but for the potential to generate embryonic stem (ES) cells. These are special pluripotent cells in a very early embryo that can develop into all the different tissues in the body (see Chapter 11). By creating personalized cloned stem cells, researchers hoped to generate genetically matched replacement organs for desperately ill patients. This idea generated a lot of debate, as some people felt it was unethical to generate human embryos for this purpose, or even to perform any research at all on human embryos or ES cells.

In 2006, a discovery by Japanese scientist Shinya Yamanaka and his colleagues changed everything. They found that adding just four proteins (all of them transcription factors) to adult cells could wind back the biological clock and turn them back into stem cells. Known as induced pluripotent stem (iPS) cells, these incredible, reprogrammed cells could then be grown into any type of cell in the body. In 2012, Yamanaka won the Nobel Prize for Physiology or Medicine (jointly with the English developmental biologist, Sir John B. Gurdon) for this breakthrough, which meant that there was no longer any need to use embryos to generate stem cells for repairing the body.

Researchers across the world are now working on ways to harness the power of iPS cells to treat or cure disease. One exciting idea is to use specially adapted 3D printers to "print" replacement donor organs made from personalized iPS cells. This kind of individualized approach is likely to be very expensive, so scientists are now working on building libraries of iPS cells that broadly match the genetic makeup of many people, which could help to bring this exciting idea into reality.

WILL HUMANS GO EXTINCT?

Modern humans are the only known species of hominin alive on Earth today. As we saw in Chapter 2, all the other relatives from our evolutionary journey – such as the Neanderthals and Denisovans – are now extinct, although their DNA still lurks within us. Animal and plant species are dying out all over the world, in many cases hastened by human activity and a changing environment, and we seem to be living in a time of accelerated extinctions. So could the same fate happen to our own species?

We know that there have been mass extinctions on Earth in the past. The last major event, which wiped out around three quarters of all the animal and plant species on the planet, including the dinosaurs, occurred around 66 million years ago, when a huge asteroid slammed into the Earth. This impact caused massive, sudden changes in the environment, making survival very difficult for all but a few lucky species. However, this mass extinction scenario is unlikely to happen again at any time in the foreseeable future, as there does not seem to be anything that big on a collision course with Earth. Neither is there likely to be a massive natural disaster, such as a super-volcano eruption, which has caused devastation in the past.

Instead, humanity is much more likely to become extinct as a result of problems we have caused for ourselves. One threat is nuclear war, which would cause devastating environmental change as well as killing and poisoning many people with radiation. Then there is disease. Thanks to over-use of antibiotics in medicine and farming, the problem of antibiotic-resistant bacteria is becoming more serious all the time. Bacteria evolve and adapt to drugs very quickly, so many scientists and doctors fear that this could eventually leave us without any options for preventing or treating bacterial infections. Not only would this cause a big public-health issue if common diseases became resistant to all drugs, but it would also mean that medical procedures that carry a risk of infection, such as surgery, would become very risky.

There is also the chance that extremely nasty, new viral infections will arise. We already know that viruses (such as swine flu) can occasionally jump from animals to humans. These can evolve to become more infectious and dangerous, leading to a future pandemic. Fortunately, some people are still likely to be resistant to viral infection (like the genetic superhero Stephen Crohn, whom we met in Chapter 7), so it is unlikely that a single virus will wipe out the entire human race.

Another major factor is climate change. The global climate changes naturally over thousands of years, but human activity – especially global warming due to carbon-dioxide production – is accelerating this pattern, changing the world in ways that humans have never seen and cannot predict. Some parts of the planet could experience intense freezing, while others heat up and become arid deserts. We might run out of fresh drinking water or land for growing food.

On a more positive note, this might not be enough to render human beings completely extinct. We have adapted to harsh and changing conditions in the past, and we can do so again as the world changes in the future. Our species teetered on the brink of extinction at several points in our early evolution, but we made it through. Maybe we will colonize another planet before humans are annihilated on Earth. However, most people would prefer to try to make the most of life down here, rather than pin all our hopes on a risky journey towards the stars.

In a few tens of thousands of years, any humans that have emerged from these impending crises will be very different from people today, in ways that we simply cannot foresee. Other forms of life will also survive, evolving in their own way in response to the changing environment. Who knows what the future holds for humans and our genes, or how we will adapt to cope with it? What we do know for sure is that if we fail to do so we will die out, like so many unsuccessful extinct species before us. Life will probably find a way – but it just might not include humans in our current form. So it really is important that we look after our genes and our planet for as long as we can.

GLOSSARY

Amino acids The chemical building blocks that make up proteins.

Bases The chemical building blocks ("letters") that make up DNA and RNA: A (adenine), C (cytosine), G (guanine) and T (thymine). Also known as nucleotides.

Chromatin DNA together with histone packaging proteins.

Chromosome A single long string of DNA. Human cells contain 23 pairs of chromosomes.

DNA methylation A chemical tag added to the base C (cytosine), creating methylcytosine (meC).

DNA polymerase An enzyme that makes a copy of DNA.

Enhancer A control switch that turns a gene on.

Enzyme A protein that does a job in a cell, such as RNA polymerase.

Epigenetic modifications Reversible chemical changes to DNA or its histone packaging proteins that are thought to convey information about gene activity.

Gene A stretch of DNA that carries the information to make a specific protein or RNA.

Genome The entire set of DNA in a cell or organism.

Germ cells Cells that make either eggs or sperm.

Histones Proteins that package DNA.

Messenger RNA (mRNA) A fully processed RNA message read from a gene, which usually encodes the instructions to make a protein.

Non-coding DNA DNA that doesn't contain protein-coding genes.

Non-coding RNA RNA transcribed from DNA that doesn't carry the instructions to make a protein.

Nucleotides See bases.

Nucleus The structure within a cell that contains DNA.

Promoter The start of a gene.

Protein A molecule made up of a long string of small building blocks called amino acids.

Ribosome A molecular "machine" that builds proteins using information encoded in messenger RNA.

RNA polymerase An enzyme that "reads" DNA and makes an RNA copy – this process is known as transcription.

Sequencing Reading the order of letters (bases) in DNA.

Transcription Making an RNA copy of the genetic information encoded within DNA.

Transcription factor A protein that helps to switch a gene on.

Translation Assembling amino acids together to make a protein, directed by messenger RNA.

I N D E X

Page numbers in **bold** refer to main entries; page numbers in *italics* refer to illustrations/photographs/captions; page numbers in ***bold italic*** refer to timelines

ACKNOWLEDGEMENTS

With thanks to Professor Chris Stringer at the Natural History Museum, London.